# The Science of the Stars

*by*

E. Walter Maunder



Double9 BOOKS

# The Science of the Stars

## by E. Walter Maunder



ISBN: 978-93-67146-47-7

**Published by**

**DOUBLE 9 BOOKS**
2/13-B, Ansari Road
Daryaganj, New Delhi – 110002
info@double9books.com
www.double9books.com
Tel. 011-40042856

# ABOUT THE AUTHOR

Edward Walter Maunder was an English astronomer who lived from 12 April 1851 to 21 March 1928. His research into sunspots and the solar magnetic cycle led to the discovery of the Maunder Minimum, which lasted from 1645 to 1715. Maunder was born in London in 1851, the youngest child of a Wesleyan Society minister. He studied at King's College London but did not graduate. To fund his studies, he obtained a job in a London bank. Maunder returned to the Royal Observatory in 1873 as a spectroscopic assistant. He married Edith Hannah Bustin in 1875, and they had six children: four sons (one died in infancy) and two girls. Following Edith's death in 1888, he met Annie Scott Dill Russell (later Annie Russell Maunder, 1868-1947), a mathematician and astronomer trained at Cambridge's Girton College, with whom he cooperated for the rest of his life, in 1890. From 1890 through 1895, she worked as a "lady computer" at the Observatory. Maunder and Russell married in 1895. Annie Maunder became one of the Royal Astronomical Society's first female members in 1916.

# CONTENTS

# CHAPTER I
# ASTRONOMY BEFORE HISTORY

The plan of the present series requires each volume to be complete in about eighty small pages. But no adequate account of the achievements of astronomy can possibly be given within limits so narrow, for so small a space would not suffice for a mere catalogue of the results which have been obtained; and in most cases the result alone would be almost meaningless unless some explanation were offered of the way in which it had been reached. All, therefore, that can be done in a work of the present size is to take the student to the starting-point of astronomy, show him the various roads of research which have opened out from it, and give a brief indication of the character and general direction of each.

That which distinguishes astronomy from all the other sciences is this: it deals with objects that we cannot touch. The heavenly bodies are beyond our reach; we cannot tamper with them, or subject them to any form of experiment; we cannot bring them into our laboratories to analyse or dissect them. We can only watch them and wait for such indications as their own movements may supply. But we are confined to this earth of ours, and they are so remote; we are so short-lived, and they are so long-enduring; that the difficulty of finding out much about them might well seem insuperable.

Yet these difficulties have been so far overcome that astronomy is the most advanced of all the sciences, the one in which our knowledge is the most definite and certain. All science rests on sight and thought, on ordered observation and reasoned deduction; but both sight and thought were earlier trained to the service of astronomy than of the other physical sciences.

It is here that the highest value of astronomy lies; in the discipline that it has afforded to man's powers of observation and reflection; and the real triumphs which it has achieved are not the bringing to light of the beauties or the sensational dimensions and distances of the heavenly bodies, but the vanquishing of difficulties which might well have seemed superhuman. The true spirit of the science can be far better exemplified by the presentation of some of these difficulties, and of the methods by which they have been overcome, than by many volumes of picturesque description or of eloquent rhapsody.

There was a time when men knew nothing of astronomy; like every other science it began from zero. But it is not possible to suppose that such a state of things lasted long, we know that there was a time when men had noticed that there were two great lights in the sky—a greater light that shone by day, a lesser light that shone by night—and there were the stars also. And this, the earliest observation of primitive astronomy, is preserved for us, expressed in the simplest possible language, in the first chapter of the first book of the sacred writings handed down to us by the Hebrews.

This observation, that there are bodies above us giving light, and that they are not all equally bright, is so simple, so inevitable, that men must have made it as soon as they possessed any mental power at all. But, once made, a number of questions must have intruded themselves: "What are these lights? Where are they? How far are they off?"

Many different answers were early given to these questions. Some were foolish; some, though intelligent, were mistaken; some, though wrong, led eventually to the discovery of the truth. Many myths, many legends, some full of beauty and interest, were invented. But in so small a book as this it is only possible to glance at those lines of thought which eventually led to the true solution.

As the greater light, the lesser light, and the stars were carefully watched, it was seen not only that they shone, but that they appeared to move; slowly, steadily, and without ceasing. The stars all moved together like a column of soldiers on the march, not altering their positions relative to each other. The lesser light, the Moon, moved with the stars, and yet at the same time among them. The greater light, the Sun, was not seen with the stars; the brightness of his presence made the day, his absence brought the night, and it was only during his absence that the stars were seen; they faded out of the sky before he came up in the morning, and did not reappear again until after he passed out of sight in the evening. But there came a time when it was realised that there were stars shining in the sky all day long as well as at night, and this discovery was one of the greatest and most important ever made, because it was the earliest discovery of something quite unseen. Men laid hold of this fact, not from the direct and immediate evidence of their senses, but from reflection and reasoning. We do not know who made this discovery, nor how long ago it was made, but from that time onward the eyes with which men looked upon nature were not only the eyes of the body, but also the eyes of the mind.

It followed from this that the Sun, like the Moon, not only moved with the general host of the stars, but also among them. If an observer looks out from any fixed station and watches the rising of some bright star, night after night, he will notice that it always appears to rise in the same place; so too with its setting. From any given observing station the direction in which any particular star is observed to rise or set is invariable.

Not so with the Sun. We are accustomed to say that the Sun rises in the east and sets in the west. But the direction in which the Sun rises in midwinter lies far to the south of the east point; the direction in which he rises in midsummer lies as far to the north. The Sun is therefore not only moving with the stars, but among them. This gradual change in the position of the Sun in the sky was noticed in many ancient nations at an early time. It is referred to in Job xxxviii. 12: "Hast thou commanded the morning since thy days; and caused the dayspring to know his place?"

And the apparent path of the Sun on one day is always parallel to its path on the days preceding and following. When, therefore, the Sun rises far to the south of east, he sets correspondingly far to the south of west, and at noon he is low down in the south. His course during the day is a short one, and the daylight is much shorter than the night, and the Sun at noon, being low down in the sky, has not his full power. The cold and darkness of winter, therefore, follows directly upon this position of the Sun. These conditions are reversed when the Sun rises in the north-east. The night is short, the daylight prolonged, and the Sun, being high in the heavens at noon, his heat is felt to the full.

Thus the movements of the Sun are directly connected with the changes of season upon the Earth. But the stars also are connected with those seasons; for if we look out immediately after it has become dark after sunset, we shall notice that the stars seen in the night of winter are only in part those seen in the nights of summer.

In the northern part of the sky there are a number of stars which are always visible whenever we look out, no matter at what time of the night nor what part of the year. If we watch throughout the whole night, we see that the whole heavens appear to be slowly turning—turning, as if all were in a single piece—and the pivot about which it is turning is high up in the northern sky. The stars, therefore, are divided into two classes. Those near this invisible pivot—the "Pole" of the Heavens, as we term it—move round it in complete circles; they never pass out of sight, but even when lowest they clear the horizon. The other stars move round the same pivot in curved paths, which are evidently parts of circles, but circles of which we do not see the whole. These stars rise on the eastern side of the heavens and set on the western, and for a greater or less space of time are lost to sight below the horizon. And some of these stars are visible at one time of the year, others at another; some being seen during the whole of the long nights of winter, others throughout the short nights of summer. This distinction again, and its connection with the change of the seasons on the earth, was observed many ages ago. It is alluded to in Job xxxviii. 32: "Canst thou lead forth the Signs of the Zodiac in their season, or canst thou guide the Bear with her train?" (R.V., Margin). The Signs of the Zodiac are taken as representing the stars which rise and set, and therefore have each their season for being "led forth," while the northern stars, which are always visible, appearing to be "guided" in their continual movement round the Pole of the sky in perfect circles, are represented by "the Bear with her train."

The changes in position of the Sun, the greater light, must have attracted attention in the very earliest ages, because these changes are so closely connected with the changes of the seasons upon the Earth, which affect men directly. The Moon, the lesser light, goes through changes of position like the Sun, but these are not of the same direct consequence to men, and probably much less notice was taken of them. But there were changes of the Moon which men could not help noticing—her changes of shape and brightness. One evening she may be seen soon after the Sun has set, as a thin arch of light, low down in the sunset sky. On the following evenings she is seen higher and higher in the sky, and the bow of light increases, until by the fourteenth day it is a perfect round. Then the Moon begins to diminish and to disappear, until, on the twenty-ninth or thirtieth day after the first observation, she is again seen in the west after sunset as a narrow crescent. This succession of changes gave men an important measure of time, and, in an age when artificial means of light were difficult to procure, moonlight was of the greatest value, and the return of the moonlit portion of the month was eagerly looked for.

These early astronomical observations were simple and obvious, and of great practical value. The day, month, and year were convenient measures of time, and the power of determining, from the observation of the Sun and of the stars, how far the year had progressed was most important to farmers, as an indication when they should plough and sow their land. Such observations had probably been made independently by many men and in many nations, but in one place a greater advance had been made. The Sun and Moon are both unmistakable, but one star is very like another, and, for the most part, individual stars can only be recognised by their positions relative to others. The stars were therefore grouped together into **Constellations** and associated with certain fancied designs, and twelve of these designs were arranged in a belt round the sky to mark the apparent path of the Sun in the course of the year, these twelve being known as the **"Signs of the Zodiac"**—the Ram, Bull, Twins, Crab, Lion, Virgin, Balance, Scorpion, Archer, Goat, Water-pourer, and Fishes. In the rest of the sky some thirty to thirty-six other groups, or constellations, were formed, the Bear being the largest and brightest of the constellations of the northern heavens.

But these ancient constellations do not cover the entire heavens; a large area in the south is untouched by them. And this fact affords an indication both of the time when and the place where the old stellar groups were designed, for the region left untouched was the region below the horizon of 40° North latitude, about 4600 years ago. It is probable, therefore, that the ancient astronomers who carried out this great work lived about 2700 B.C., and in North latitude 37° or 38°. The indication is only rough, but the amount of uncertainty is not very large; the constellations must be at least 4000 years old, they cannot be more than 5000.

All this was done by prehistoric astronomers; though no record of the actual carrying out of the work and no names of the men who did it have come down to us. But it is clear from the fact that the Signs of the Zodiac are arranged so as to mark out the annual path of the Sun, and that they are twelve in number—there being twelve months in the year—that those who designed the constellations already knew that there are stars shining near the Sun in full daylight, and that they had worked out some means for determining what stars the Sun is near at any given time.

Another great discovery of which the date and the maker are equally unknown is referred to in only one of the ancient records available to us. It was seen that all along the eastern horizon, from north to south, stars rise, and all along the western horizon, from north to south, stars set. That is what was seen; it was the fact observed. There is no hindrance anywhere to the movement of the stars—they have a free passage under the Earth; the Earth is unsupported in space. That is what was *thought*; it was the inference drawn. Or, as it is written in Job xxvi. 7, "He (God) stretcheth out the north over empty space, and hangeth the earth upon nothing."

The Earth therefore floats unsupported in the centre of an immense star-spangled sphere. And what is the shape of the Earth? The natural and correct inference is that it is spherical, and we find in some of the early Greek writers the arguments which establish this inference as clearly set forth as they would be to-day. The same inference followed, moreover, from the observation of a simple fact, namely, that the stars as observed from any particular place all make the same angle with the horizon as they rise in the east, and all set at the same angle with it in the west; but if we go northward, we find that angle steadily decreasing; if we go southward, we find it increasing. But if the Earth is round like a globe, then it must have a definite size, and that size can be measured. The discoveries noted above were made by men whose names have been lost, but the name of the first person whom we know to have measured the size of the Earth was ERATOSTHENES. He found that the Sun was directly overhead at noon at midsummer at Syene (the modern Assouan), in Egypt, but was 7° south of the "zenith"—the point overhead—at Alexandria, and from this he computed the Earth to be 250,000 stadia (a stadium = 606 feet) in circumference.

Another consequence of the careful watch upon the stars was the discovery that five of them were planets; "wandering" stars; they did not move all in one piece with the rest of the celestial host. In this they resemble the Sun and Moon, and they further resemble the Moon in that, though too small for any change of shape to be detected, they change in brightness from time to time. But their movements are more complicated than those of the other heavenly bodies. The Sun moves a little slower than the stars, and so seems to travel amongst them from west to east; the Moon moves much slower than the stars, so her motion from west to east is more pronounced than that of the Sun. But the five planets sometimes move slower than the stars, sometimes quicker, and sometimes at the same rate. Two of the five, which we now know as Mercury and Venus, never move far from the Sun, sometimes being seen in the east before he rises in the morning, and sometimes in the west after he has set in the evening. Mercury is the closer to the Sun, and moves more quickly; Venus goes through much the greater changes of brightness. Jupiter and Saturn move nearly at the same average rate as the stars, Saturn taking about thirteen days more than a year to come again to the point of the sky opposite to the Sun, and Jupiter about thirty-four days. Mars, the fifth planet, takes two years and fifty days to accomplish the same journey.

These planetary movements were not, like those of the Sun and Moon and stars, of great and obvious consequence to men. It was important to men to know when they would have moonlight nights, to know when the successive seasons of the year would return. But it was no help to men to know when Venus was at her brightest more than when she was invisible. She gave them no useful light, and she and her companion planets returned at no definite seasons. Nevertheless, men began to make ordered observations of the planets—observations that required much more patience and perseverance than those of the other celestial lights. And they set themselves with the greatest ingenuity to unravel the secret of their complicated and seemingly capricious movements.

This was a yet higher development than anything that had gone before, for men were devoting time, trouble, and patient thought, for long series of years, to an inquiry which did not promise to bring them any profit or advantage. Yet the profit which it actually did bring was of the highest order. It developed men's mental powers; it led to the devising of instruments of precision for the observations; it led to the foundation of mathematics, and thus lay at the root of all our modern mechanical progress. It brought out, in a higher degree, ordered observation and ordered thought.

# CHAPTER II
# ASTRONOMY BEFORE THE TELESCOPE

There was thus a real science of astronomy before we have any history of it. Some important discoveries had been made, and the first step had been taken towards cataloguing the fixed stars. It was certainly known to some of the students of the heavens, though perhaps only to a few, that the Earth was a sphere, freely suspended in space, and surrounded on all sides by the starry heavens, amongst which moved the Sun, Moon, and the five planets. The general character of the Sun's movement was also known; namely, that he not only moved day by day from east to west, as the stars do, but also had a second motion inclined at an angle to the first, and in the opposite direction, which he accomplished in the course of a year.

To this sum of knowledge, no doubt, several nations had contributed. We do not know to what race we owe the constellations, but there are evidences of an elementary acquaintance with astronomy on the part of the Chinese, the Babylonians, the Egyptians, and the Jews. But in the second stage of the development of the science the entire credit for the progress made belongs to the Greeks.

The Greeks, as a race, appear to have been very little apt at originating ideas, but they possessed, beyond all other races, the power of developing and perfecting crude ideas which they had obtained from other sources, and when once their attention was drawn to the movements of the heavenly bodies, they devoted themselves with striking ingenuity and success to devising theories to account for the appearances presented, to working out methods of computation, and, last, to devising instruments for observing the places of the luminaries in which they were interested.

In the brief space available it is only possible to refer to two or three of the men whose commanding intellects did so much to help on the development of the science. EUDOXUS of Knidus, in Asia Minor (408-355 B.C.), was, so far as we know, the first to attempt to represent the movements of the heavenly bodies by a simple mathematical process. His root idea was something like this. The Earth was in the centre of the universe, and it was surrounded, at a great distance from us, by a number of invisible transparent shells, or spheres. Each of these spheres rotated with perfect uniformity, though the speed of rotation differed for different spheres. One sphere carried the stars, and rotated from east to west in about 23 h. 56 m. The Sun was carried by another sphere, which rotated from west to east in a year, but the pivots, or poles, of this sphere were carried by a second, rotating exactly like the sphere of the stars. This explained how it is that the ecliptic—that is to say, the apparent path of the Sun amongst the stars—is inclined 23-½° to the equator of the sky, so that the Sun is 23-½° north of the equator at midsummer and 23-½° south of the equator at midwinter, for the poles of the sphere peculiar to the Sun were supposed to be 23-½° from the poles of the sphere peculiar to the stars. Then the Moon had three spheres; that which actually carried the Moon having its poles 5° from the poles of the sphere peculiar to the Sun. These poles were carried by a sphere placed like the sphere of the Sun, but rotating in 27 days; and this, again, had its poles in the sphere of the stars. The sphere carrying the Moon afforded the explanation of the wavy motion of the Moon to and fro across the ecliptic in the course of a month, for at one time in the month the Moon is 5° north of the ecliptic, at another time 5° south. The motions of the planets were more difficult to represent, because they not only have a general daily motion from east to west, like the stars, and a general motion from west to east along the ecliptic, like the Sun and Moon, but from time to time they turn back on their course in the ecliptic, and "retrograde." But the introduction of a third and fourth sphere enabled the motions of most of the planets to be fairly represented. There were thus twenty-seven spheres in all—four for each of the five planets, three for the Moon, three for the Sun (including one not mentioned in the foregoing summary), and one for the stars. These spheres were not, however, supposed to be solid structures really existing; the theory was simply a means for representing the observed motions of the heavenly bodies by computations based upon a series of uniform movements in concentric circles.

But this assumption that each heavenly body moves in its path at a uniform rate was soon seen to be contrary to fact. A reference to the almanac will show at once that the Sun's movement is not uniform. Thus for the year 1910-11 the solstices and equinoxes fell as given on the next page:

| *Epoch* | *Time* | *Interval* |
|---|---|---|
| Winter Solstice | 1910 Dec. 22 d. 5 h. 12 m. P.M. | 89 d. 0 h. 42 m. |
| Spring Equinox | 1911 Mar. 21 " 5 " 54 " P.M. | 92 " 19 " 41 " |
| Summer Solstice | 1911 June 22 " 1 " 35 " P.M. | 93 " 14 " 43 " |
| Autumn Equinox | 1911 Sept. 24 " 4 " 18 " A.M. | 89 " 18 " 36 " |
| Winter Solstice | 1911 Dec. 22 " 10 " 54 " P.M. | |

so that the winter half of the year is shorter than the summer half; the Sun moves more quickly over the half of its orbit which is south of the equator than over the half which is north of it.

The motion of the Moon is more irregular still, as we can see by taking out from the almanac the times of new and full moon:

| *New Moon* | *Interval to Full Moon* |
|---|---|
| Dec. 1910 1 d. 9 h. 10.7 m. P.M. | 14 d. 13 h. 54.4 m. |
| " " 31 " 4 " 21.2 " P.M. | 14 " 6 " 4.8 " |
| Jan. 1911 30 " 9 " 44.7 " A.M. | 14 " 0 " 52.8 " |
| March " 1 " 0 " 31.1 " A.M. | 13 " 23 " 27.4 " |
| " " 30 " 0 " 37.8 " P.M. | 14 " 1 " 58.8 " |
| April " 28 " 10 " 25.0 " P.M. | 14 " 7 " 44.7 " |
| May " 28 " 6 " 24.4 " A.M. | 14 " 15 " 26.3 " |
| June " 26 " 1 " 19.7 " P.M. | 14 " 23 " 33.7 " |
| July " 25 " 8 " 12.0 " P.M. | 15 " 6 " 42.7 " |
| Aug. " 24 " 4 " 14.3 " A.M. | 15 " 11 " 42.4 " |
| Sept. " 22 " 2 " 37.4 " P.M. | 15 " 13 " 33.7 " |
| Oct. " 22 " 4 " 9.3 " A.M. | 15 " 11 " 38.8 " |
| Nov. " 20 " 8 " 49.4 " P.M. | 15 " 6 " 2.5 " |
| Dec. " 20 " 3 " 40.3 " P.M. | 14 " 21 " 49.4 " |

| *Full Moon* | *Interval to New Moon* |
|---|---|
| Dec. 1910 16 d 11 h. 5.1 m. A.M. | 15 d. 5 h. 16.1 m. |
| Jan. 1911 14 " 10 " 26.0 " P.M. | 15 " 11 " 18.7 " |
| Feb. " 13 " 10 " 37.5 " A.M. | 15 " 13 " 53.6 " |
| March " 14 " 11 " 58.5 " P.M. | 15 " 12 " 39.3 " |
| April " 13 " 2 " 36.6 " P.M. | 15 " 7 " 48.4 " |
| May " 13 " 6 " 9.7 " A.M. | 15 " 0 " 14.7 " |
| June " 11 " 9 " 50.7 " P.M. | 14 " 15 " 29.0 " |
| July " 11 " 0 " 53.4 " P.M. | 14 " 7 " 18.6 " |
| Aug. " 10 " 2 " 54.7 " A.M. | 14 " 1 " 19.6 " |
| Sept. " 8 " 3 " 56.7 " P.M. | 13 " 22 " 40.7 " |
| Oct. " 8 " 4 " 11.1 " A.M. | 13 " 23 " 58.2 " |
| Nov. " 6 " 3 " 48.1 " P.M. | 14 " 5 " 1.3 " |
| Dec. " 6 " 2 " 51.9 " A.M. | 14 " 12 " 48.4 " |
| Jan. 1912 4 " 1 " 99.7 " P.M. | 14 " 21 " 40.3 " |

The astronomer who dealt with this difficulty was HIPPARCHUS (about 190-120 B.C.), who was born at Nicæa, in Bithynia, but made most of his astronomical observations in Rhodes. He attempted to explain these irregularities in the motions of the Sun and Moon by supposing that though they really moved uniformly in their orbits, yet the centre of their orbits was not the centre of the Earth, but was situated a little distance from it. This point was called "**the excentric**," and the line from the excentric to the Earth was called "**the line of apsides**."

But when he tried to deal with the movements of the planets, he found that there were not enough good observations available for him to build up any satisfactory theory. He therefore devoted himself to the work of making systematic determinations of the places of the planets that he might put his successors in a better position to deal with the problem than he was. His great successor was CLAUDIUS PTOLEMY of Alexandria, who carried the work of astronomical observation from about A.D. 127 to 150. He was, however, much greater as a mathematician than as an observer, and he worked out a very elaborate scheme, by which he was able to represent the motions of the planets with considerable accuracy. The system was an extremely complex one, but its principle may be represented as follows: If we suppose that a planet is moving round the Earth in a circle at a uniform rate, and we tried to compute the place of the planet on this assumption for regular intervals of time, we should find that the planet gradually got further and further away from the predicted place. Then after a certain time the error would reach a maximum, and begin to diminish, until the error vanished and the planet was in the predicted place at the proper time. The error would then begin to fall in the opposite direction, and would increase as before to a maximum, subsequently diminishing again to zero. This state of things might be met by supposing that the planet was not itself carried by the circle round the earth, but by an **epicycle**—*i.e.* a circle travelling upon the first circle—and by judiciously choosing the size of the epicycle and the time of revolution the bulk of the errors in the planet's place might be represented. But still there would be smaller errors going through their own period, and these, again, would have to be met by imagining that the first epicycle carried a second, and it might be that the second carried a third, and so on.

The Ptolemaic system was more complicated than this brief summary would suggest, but it is not possible here to do more than indicate the general principles upon which it was founded, and the numerous other systems or modifications of them produced in the five centuries from Eudoxus to

Ptolemy must be left unnoticed. The point to be borne in mind is that one fundamental assumption underlay them all, an assumption fundamental to all science—the assumption that like causes must always produce like effects. It was apparent to the ancient astronomers that the stars—that is to say, the great majority of the heavenly bodies—do move round the Earth in circles, and with a perfect uniformity of motion, and it seemed inevitable that, if one body moved round another, it should thus move. For if the revolving body came nearer to the centre at one time and receded at another, if it moved faster at one time and slower at another, then, the cause remaining the same, the effect seemed to be different. Any complexity introduced by superposing one epicycle upon another seemed preferable to abandoning this great fundamental principle of the perfect uniformity of the actings of Nature.

For more than 1300 years the Ptolemaic system remained without serious challenge, and the next great name that it is necessary to notice is that of COPERNICUS (1473-1543). Copernicus was a canon of Frauenburg, and led the quiet, retired life of a student. The great work which made him immortal, *De Revolutionibus*, was the result of many years' meditation and work, and was not printed until he was on his deathbed. In this work Copernicus showed that he was one of those great thinkers who are able to look beyond the mere appearance of things and to grasp the reality of the unseen. Copernicus realised that the appearance would be just the same whether the whole starry vault rotated every twenty-four hours round an immovable Earth from east to west or the Earth rotated from west to east in the midst of the starry sphere; and, as the stars are at an immeasurable distance, the latter conception was much the simpler. Extending the idea of the Earth's motion further, the supposition that, instead of the Sun revolving round a fixed Earth in a year, the Earth revolved round a fixed Sun, made at once an immense simplification in the planetary motions. The reason became obvious why Mercury and Venus were seen first on one side of the Sun and then on the other, and why neither of them could move very far from the Sun; their orbits were within the orbit of the Earth. The stationary points and retrogressions of the planets were also explained; for, as the Earth was a planet, and as the planets moved in orbits of different sizes, the outer planets taking a longer time to complete a revolution than the inner, it followed, of necessity, that the Earth in her motion would from time to time be passed by the two inner planets, and would overtake the three outer. The chief of the Ptolemaic epicycles were done away with, and all the planets moved continuously in the same direction round the Sun. But no planet's motion could be represented by uniform motion in a single circle, and Copernicus had still to make use of systems of epicycles to account for the deviations from regularity in the planetary motions round the Sun. The Earth having been abandoned as the centre of the universe, a further sacrifice had to be made: the principle of uniform motion in a circle, which had seemed so necessary and inevitable, had also to be given up.

For the time came when the instruments for measuring the positions of the stars and planets had been much improved, largely due to TYCHO BRAHE (1546-1601), a Dane of noble birth, who was the keenest and most careful observer that astronomy had yet produced. His observations enabled his friend and pupil, JOHANN KEPLER, (1571-1630), to subject the planetary movements to a far more searching examination than had yet been attempted, and he discovered that the Sun is in the plane of the orbit of each of the planets, and also in its **line of apsides**—that is to say, the line joining the two points of the orbit which are respectively nearest and furthest from the Sun. Copernicus had not been aware of either of these two relations, but their discovery greatly strengthened the Copernican theory.

Then for many years Kepler tried one expedient after another in order to find a combination of circular motions which would satisfy the problem before him, until at length he was led to discard the circle and try a different curve—the oval or ellipse. Now the property of a circle is that every point of it is situated at the same distance from the centre, but in an ellipse there are two points within it, the "foci," and the sum of the distances of any point on the circumference from these two foci is constant. If the two foci are at a great distance from each other, then the ellipse is very long and narrow; if the foci are close together, the ellipse differs very little from a circle; and if we imagine that the two foci actually coincide, the ellipse becomes a circle. When Kepler tried motion in an ellipse instead of motion in a circle, he found that it represented correctly the motions of all the planets without any need for epicycles, and that in each case the Sun occupied one of the foci. And though the planet did not move at a uniform speed in the ellipse, yet its motion was governed by a uniform law, for the straight line joining the planet to the Sun, the "**radius vector**," passed over equal areas of space in equal periods of time.

These two discoveries are known as Kepler's First and Second Laws. His Third Law connects all the planets together. It was known that the outer planets not only take longer to revolve round the Sun than the inner, but that their actual motion in space is slower, and Kepler found that this actual speed of motion is inversely as the square root of its distance from the Sun;

or, if the square of the speed of a planet be multiplied by its distance from the Sun, we get the same result in each case. This is usually expressed by saying that the cube of the distance is proportional to the square of the time of revolution. Thus the varying rate of motion of each planet in its orbit is not only subject to a single law, but the very different speeds of the different planets are also all subject to a law that is the same for all.

Thus the whole of the complicated machinery of Ptolemy had been reduced to three simple laws, which at the same time represented the facts of observation much better than any possible development of the Ptolemaic mechanism. On his discovery of his third law Kepler had written: "The book is written to be read either now or by posterity—I care not which; it may well wait a century for a reader, as God has waited 6000 years for an observer." Twelve years after his death, on Christmas Day 1642 (old style), near Grantham, in Lincolnshire, the predestined "reader" was born. The inner meaning of Kepler's three laws was brought to light by ISAAC NEWTON.

# CHAPTER III
# THE LAW OF GRAVITATION

The fundamental thought which, recognised or not, had lain at the root of the Ptolemaic system, as indeed it lies at the root of all science, was that "like causes must always produce like effects." Upon this principle there seemed to the ancient astronomers no escape from the inference that each planet must move at a uniform speed in a circle round its centre of motion. For, if there be any force tending to alter the distance of the planet from that centre, it seemed inevitable that sooner or later it should either reach that centre or be indefinitely removed from it. If there be no such force, then the planet's distance from that centre must remain invariable, and if it move at all, it must move in a circle; move uniformly, because there is no force either to hasten or retard it. Uniform motion in a circle seemed a necessity of nature.

But all this system, logical and inevitable as it had once seemed, had gone down before the assault of observed facts. The great example of uniform circular motion had been the daily revolution of the star sphere; but this was now seen to be only apparent, the result of the rotation of the Earth. The planets revolved round the Sun, but the Sun was not in the centre of their motion; they moved, not in circles, but in ellipses; not at a uniform speed, but at a speed which diminished with the increase of their distance from the Sun. There was need, therefore, for an entire revision of the principles upon which motion was supposed to take place.

The mistake of the ancients had been that they supposed that continued motion demanded fresh applications of force. They noticed that a ball, set rolling, sooner or later came to a stop; that a pendulum, set swinging, might swing for a good time, but eventually came to rest; and, as the forces that were checking the motion—that is to say, the friction exercised by the ground, the atmosphere, and the like—did not obtrude themselves, they were overlooked.

Newton brought out into clear statement the true conditions of motion. A body once moving, if acted upon by no force whatsoever, must continue to move forward in a straight line at exactly the same speed, and that for ever. It does not require any maintaining force to keep it going. If any change in its speed or in its direction takes place, that change must be due to the introduction of some further force.

This principle, that, if no force acts on a body in motion, it will continue to move uniformly in a straight line, is Newton's First Law of Motion. His Second lays it down that, if force acts on a body, it produces a change of motion proportionate to the force applied, and in the same direction. And the Third Law states that when one body exerts force upon another, that second body reacts with equal force upon the first. The problem of the motions of the planets was, therefore, not what kept them moving, but what made them deviate from motion in a straight line, and deviate by different amounts.

It was quite clear, from the work of Kepler, that the force deflecting the planets from uniform motion in a straight line lay in the Sun. The facts that the Sun lay in the plane of the orbits of all the planets, that the Sun was in one of the foci of each of the planetary ellipses, that the straight line joining the Sun and planet moved for each planet over equal areas in equal periods of time, established this fact clearly. But the amount of deflection was very different for different planets. Thus the orbit of Mercury is much smaller than that of the Earth, and is travelled over in a much shorter time, so that the distance by which Mercury is deflected in a course of an hour from movement in a straight line is much greater than that by which the Earth is deflected in the same time, Mercury falling towards the Sun by about 159 miles, whilst the fall of the Earth is only about 23.9 miles. The force drawing Mercury towards the Sun is therefore 6.66 times that drawing the Earth, but 6.66 is the square of 2.58, and the Earth is 2.58 times as far from the Sun as Mercury. Similarly, the fall in an hour of Jupiter towards the Sun is about 0.88 miles, so that the force drawing the Earth is 27 times that drawing Jupiter towards the Sun. But 27 is the square of 5.2, and Jupiter is 5.2 times as far from the Sun as the Earth. Similarly with the other planets. The force, therefore, which deflects the planets from motion in a straight line, and compels them to move round the Sun, is one which varies inversely as the square of the distance.

But the Sun is not the only attracting body of which we know. The old Ptolemaic system was correct to a small extent; the Earth is the centre of motion for the Moon, which revolves round it at a mean distance of 238,800 miles, and in a period of 27 d. 7 h. 43 m. Hence the circumference of her orbit is 1,500,450 miles, and the length of the straight line which she would travel in one second of time, if not deflected by the Earth, is 2828 feet. In this distance the deviation of a circle from a straight line is one inch divided by 18.66. But we know from experiment that a stone let fall from a height of

193 inches above the Earth's surface will reach the ground in exactly one second of time. The force drawing the stone to the Earth, therefore, is 193 x 18.66; *i.e.* 3601 times as great as that drawing the Moon. But the stone is only 1/330 of a mile from the Earth's surface, while the Moon is 238,800 miles away—more than 78 million times as far. The force, therefore, would seem not to be diminished in the proportion that the distance is increased—much less in the proportion of its square.

But Newton proved that a sphere of uniform density, or made up of any number of concentric shells of uniform density, attracted a body outside itself, just as if its entire mass was concentrated at its centre. The distance of the stone from the Earth must therefore be measured, not from the Earth's surface, but from its centre; in other words, we must consider the stone as being distant from the Earth, not some 16 feet, but 3963 miles. This is very nearly one-sixtieth of the Moon's distance, and the square of 60 is 3600. The Earth's pull upon the Moon, therefore, is almost exactly in the inverse square of the distance as compared with its pull on the stone.

Kepler's book had found its "reader." His three laws were but three particular aspects of Newton's great discovery that the planets moved under the influence of a force, lodged in the Sun, which varied inversely as the square of their distances from it. But Newton's work went far beyond this, for he showed that the same law governed the motion of the Moon round the Earth and the motions of the satellites revolving round the different planets, and also governed the fall of bodies upon the Earth itself. It was universal throughout the solar system. The law, therefore, is stated as of universal application. "Every particle of matter in the universe attracts every other particle with a force varying inversely as the square of the distance between them, and directly as the product of the masses of the two particles." And Newton further proved that if a body, projected in free space and moving with any velocity, became subject to a central force acting, like gravitation, inversely as the square of the distance, it must revolve in an ellipse, or in a closely allied curve.

These curves are what are known as the "**conic sections**"—that is, they are the curves found when a cone is cut across in different directions. Their relation to each other may be illustrated thus. If we have a very powerful light emerging from a minute hole, then, if we place a screen in the path of the beam of light, and exactly at right angles to its axis, the light falling on the screen will fill an exact circle. If we turn the screen so as to be inclined to the axis of the beam, the circle will lengthen out in one direction, and will become an ellipse. If we turn the screen still further, the ellipse will lengthen and lengthen, until at last, when the screen has become parallel to one of the edges of the beam of light, the ellipse will only have one end; the other will be lost. For it is clear that that edge of the beam of light which is parallel to the screen can never meet it. The curve now shown on the screen is called a **parabola**, and if the screen is turned further yet, the boundaries of the light falling upon it become divergent, and we have a fourth curve, the **hyperbola**. Bodies moving under the influence of gravitation can move in any of these curves, but only the circle and ellipse are closed orbits. A particle moving in a parabola or hyperbola can only make one approach to its attracting body; after such approach it continually recedes from it. As the circle and parabola are only the two extreme forms of an ellipse, the two foci being at the same point for the circle and at an infinite distance apart for the parabola, we may regard all orbits under gravitation as being ellipses of one form or another.

From his great demonstration of the law of gravitation, Newton went on to apply it in many directions. He showed that the Earth could not be truly spherical in shape, but that there must be a flattening of its poles. He showed also that the Moon, which is exposed to the attractions both of the Earth and of the Sun, and, to a sensible extent, of some of the other planets, must show irregularities in her motion, which at that time had not been noticed. The Moon's orbit is inclined to that of the Earth, cutting its plane in two opposite points, called the "**nodes**." It had long been observed that the position of the nodes travelled round the ecliptic once in about nineteen years. Newton was able to show that this was a consequence of the Sun's attraction upon the Moon. And he further made a particular application of the principle thus brought out, for, the Earth not being a true sphere, but flattened at the poles and bulging at the equator, the equatorial belt might be regarded as a compact ring of satellites revolving round the Earth's equator. This, therefore, would tend to retrograde precisely as the nodes of a single satellite would, so that the axis of the equatorial belt of the Earth—in other words, the axis of the Earth—must revolve round the pole of the ecliptic. Consequently the pole of the heavens appears to move amongst the stars, and the point where the celestial equator crosses the equator necessarily moves with it. This is what we know as the "**Precession of the Equinoxes**," and it is from our knowledge of the fact and the amount of precession that we are able to determine roughly the date when the first great work of astronomical observation was accomplished, namely, the grouping of the stars into constellations by the astronomers of the prehistoric age.

The publication of Newton's great work, the *Principia* (*The Mathematical Principles of Natural Philosophy*), in which he developed the Laws of Motion, the significance of Kepler's Three Planetary Laws, and the Law of Universal Gravitation, took place in 1687, and was due to his friend EDMUND HALLEY, to whom he had confided many of his results. That he was the means of securing the publication of the *Principia* is Halley's highest claim to the gratitude of posterity, but his own work in the field which Newton had opened was of great importance. Newton had treated **comets** as moving in parabolic orbits, and Halley, collecting all the observations of comets that were available to him, worked out the particulars of their orbits on this assumption, and found that the elements of three were very closely similar, and that the interval between their appearances was nearly the same, the comets having been seen in 1531, 1607, and 1682. On further consulting old records he found that comets had been observed in 1456, 1378, and 1301. He concluded that these were different appearances of the same object, and predicted that it would be seen again in 1758, or, according to a later and more careful computation, in 1759. As the time for its return drew near, CLAIRAUT computed with the utmost care the retardation which would be caused to the comet by the attractions of Jupiter and Saturn. The comet made its predicted nearest approach to the Sun on March 13, 1759, just one month earlier than Clairaut had computed. But in its next return, in 1835, the computations effected by PONTÉCOULANT were only two days in error, so carefully had the comet been followed during its unseen journey to the confines of the solar system and back again, during a period of seventy-five years. Pontécoulant's exploit was outdone at the next return by Drs. COWELL and CROMMELIN, of Greenwich Observatory, who not only computed the time of its perihelion passage—that is to say, its nearest approach to the Sun—for April 16, 1910, but followed the comet back in its wanderings during all its returns to the year 240 B.C. Halley's Comet, therefore, was the first comet that was known to travel in a closed orbit and to return to the neighbourhood of the Sun. Not a few small or telescopic comets are now known to be "periodic," but Halley's is the only one which has made a figure to the naked eye. Notices of it occur not a few times in history; it was the comet "like a flaming sword" which Josephus described as having been seen over Jerusalem not very long before the destruction by Titus. It was also the comet seen in the spring of the year when William the Conqueror invaded England, and was skilfully used by that leader as an omen of his coming victory.

The law of gravitation had therefore enabled men to recognise in Halley's Comet an addition to the number of the primary bodies in the solar system—the first addition that had been made since prehistoric times. On March 13, 1781, Sir WILLIAM HERSCHEL detected a new object, which he at first supposed to be a comet, but afterwards recognised as a planet far beyond the orbit of Saturn. This planet, to which the name of Uranus was finally given, had a mean distance from the Sun nineteen times that of the

Earth, and a diameter four times as great. This was a second addition to the solar system, but it was a discovery by sight, not by deduction.

The first day of the nineteenth century, January 1, 1801, was signalised by the discovery of a small planet by PIAZZI. The new object was lost for a time, but it was redetected on December 31 of the same year. This planet lay between the orbits of Mars and Jupiter—a region in which many hundreds of other small bodies have since been found. The first of these "**minor planets**" was called Ceres; the next three to be discovered are known as Pallas, Juno, and Vesta. Beside these four, two others are of special interest: one, Eros, which comes nearer the Sun than the orbit of Mars—indeed at some oppositions it approaches the Earth within 13,000,000 miles, and is therefore, next to the Moon, our nearest neighbour in space; the other, Achilles, moves at a distance from the Sun equal to that of Jupiter.

Ceres is much the largest of all the minor planets; indeed is larger than all the others put together. Yet the Earth exceeds Ceres 4000 times in volume, and 7000 times in mass, and the entire swarm of minor planets, all put together, would not equal in total volume one-fiftieth part of the Moon.

The search for these small bodies rendered it necessary that much fuller and more accurate maps of the stars should be made than had hitherto been attempted, and this had an important bearing on the next great event in the development of gravitational astronomy.

The movements of Uranus soon gave rise to difficulties. It was found impossible, satisfactorily, to reconcile the earlier and later observations, and in the tables of Uranus, published by BOUVARD in 1821, the earlier observations were rejected. But the discrepancies between the observed and calculated places for the planet soon began to reappear and quickly increase, and the suggestion was made that these discrepancies were due to an attraction exercised by some planet as yet unknown. Thus Mrs. Somerville in a little book on the connection of the physical sciences, published in 1836, wrote, "Possibly it (that is, Uranus) may be subject to disturbances from some unseen planet revolving about the Sun beyond the present boundaries of our system. If, after the lapse of years, the tables formed from a combination of numerous observations should still be inadequate to represent the motions of Uranus, the discrepancies may reveal the existence, nay, even the mass and orbit of a body placed for ever beyond the sphere of vision." In 1843 JOHN C. ADAMS, who had just graduated as Senior Wrangler at Cambridge, proceeded to attack the problem of determining the position, orbit, and mass of the unknown body by which on this assumption Uranus was disturbed, from the irregularities evident in the motion of that planet. The problem was one of extraordinary intricacy, but by September 1845 Adams had obtained a first solution, which, he submitted to AIRY, the Astronomer Royal. As, however, he neglected to reply to some inquiries made by Airy, no search for the new planet was instituted in England until the results of a new and independent worker had been published. The same problem had been attacked by a well-known and very gifted French mathematician, U. J. J. LEVERRIER, and in June 1846 he published his position for the unseen planet, which proved to be in close accord with that which Adams had furnished to Airy nine months before. On this Airy stirred up Challis, the Director of the Cambridge Observatory, which then possessed the most powerful telescope in England, to search for the planet, and Challis commenced to make charts, which included more than 3000 stars, in order to make sure that the stranger should not escape his net. Leverrier, on the other hand, communicated his result to the Berlin Observatory, where they had just received some of the star charts prepared by Dr. Bremiker in connection with the search for minor planets. The Berlin observer, Dr. Galle, had therefore nothing to do but to compare the stars in the field, upon which he turned his telescope, with those shown on the chart; a star not in the chart would probably be the desired stranger. He found it, therefore, on the very first evening, September 23, 1846, within less than four diameters of the Moon of the predicted place. The same object had been observed by Challis at Cambridge on August 4 and 12, but he was deferring the reduction of his observations until he had completed his scrutiny of the zone, and hence had not recognised it as different from an ordinary star.

This discovery of the planet now known as Neptune, which had been disturbing the movement of Uranus, has rightly been regarded as the most brilliant triumph of gravitational astronomy. It was the legitimate crown of that long intellectual struggle which had commenced more than 2000 years earlier, when the first Greek astronomers set themselves to unravel the apparently aimless wanderings of the planets in the assured faith that they would find them obedient unto law. But of what use was all this effort? What is the good of astronomy? The question is often asked, but it is the question of ignorance. The use of astronomy is the development which it has given to the intellectual powers of man. Directly the problem of the planetary motions was first attempted, it became necessary to initiate mathematical processes in order to deal with it, and the necessity for the continued development of mathematics has been felt in the same connection right down to the present day. When the Greek astronomers first began their inquiries into the planetary movements they hoped for no material gain, and they received none. They laboured; we have entered into their labours. But the whole of our vast advances in mechanical and engineering science—advances which more than anything else differentiate this our present age from all those which have preceded it—are built upon our command of mathematics and our knowledge of the laws of motion—a command and a knowledge which we owe directly to their persevering attempts to advance the science of astronomy, and to follow after knowledge, not for any material rewards which she had to offer, but for her own sake.

# CHAPTER IV
# ASTRONOMICAL MEASUREMENTS

The old proverb has it that "Science is measurement," and of none of the sciences is this so true as of the science of astronomy. Indeed the measurement of time by observation of the movements of the heavenly bodies was the beginning of astronomy. The movement of the Sun gave the day, which was reckoned to begin either at sunrise or at sunset. The changes of the Moon gave the month, and in many languages the root meaning of the word for *Moon* is "measurer." The apparent movement of the Sun amongst the stars gave a yet longer division of time, the year, which could be determined in a number of different ways, either from the Sun alone, or from the Sun together with the stars. A very simple and ancient form of instrument for measuring this movement of the Sun was the obelisk, a pillar with a pointed top set up on a level pavement. Such obelisks were common in Egypt, and one of the most celebrated, known as Cleopatra's Needle, now stands on the Thames Embankment. As the Sun moved in the sky, the shadow of the pillar moved on the pavement, and midday, or noon, was marked when the shadow was shortest. The length of the shadow at noon varied from day to day; it was shortest at mid-summer, and longest at midwinter, *i.e.* at the summer and winter solstices. Twice in the year the shadow of the pillar pointed due west at sunrise, and due east at sunset—that is to say, the shadow at the beginning of the day was in the same straight line as at its end. These two days marked the two equinoxes of spring and autumn.

The obelisk was a simple means of measuring the height and position of the Sun, but it had its drawbacks. The length of the shadow and its direction did not vary by equal amounts in equal times, and if the pavement upon which the shadow fell was divided by marks corresponding to equal intervals of time for one day of the year, the marks did not serve for all other days.

But if for the pillar a triangular wall was substituted—a wall rising from the pavement at the south and sloping up towards the north at such an angle that it seemed to point to the invisible pivot of the heavens, round which all the stars appeared to revolve—then the shadow of the wall moved on the pavement in the same manner every day, and the pavement if marked to show the hours for one day would show them for any day. The sundials still often found in the gardens of country houses or in churchyards are miniatures of such an instrument.

But the Greek astronomers devised other and better methods for determining the positions of the heavenly bodies. Obelisks or dials were of use only with the Sun and Moon which cast shadows. To determine the position of a star, "sights" like those of a rifle were employed, and these were fixed to circles which were carefully divided, generally into 360 "degrees." As there are 365 days in a year, and as the Sun makes a complete circuit of the Zodiac in this time, it moves very nearly a degree in a day. The twelve Signs of the Zodiac are therefore each 30° in length, and each takes on the average a double-hour to rise or set. While the Sun and Moon are each about half a degree in diameter, *i.e.* about one-sixtieth of the length of a Sign, and therefore take a double-minute to rise or set. Each degree of a circle is therefore divided into 60 minutes, and each minute may be divided into 60 seconds.

As the Sun or Moon are each about half a degree, or, more exactly, 32 minutes in diameter, it is clear that, so long as astronomical observations were made by the unaided sight, a minute of arc (written 1′) was the smallest division of the circle that could be used. A cord or wire can indeed be detected when seen projected against a moderately bright background if its thickness is a second of arc (written 1″)—a sixtieth of a minute—but the wire is merely perceived, not properly defined.

Tycho Brahe had achieved the utmost that could be done by the naked eye, and it was the certainty that he could not have made a mistake in an observation in the place of the planet Mars amounting to as much as 8 minutes of arc—that is to say, of a quarter the apparent diameter of the Moon—that made Kepler finally give up all attempts to explain the planetary movements on the doctrine of circular orbits and to try movements in an ellipse. But a contemporary of Kepler, as gifted as he was himself, but in a different direction, was the means of increasing the observing power of the astronomer. GALILEO GALILEI (1564-1642), of a noble Florentine family, was appointed Lecturer in Mathematics at the University of Pisa. Here he soon distinguished himself by his originality of thought, and the ingenuity and decisiveness of his experiments. Up to that time it had been taught that of two bodies the heavier would fall to the ground more quickly than the lighter. Galileo let fall a 100-lb. weight and a 1-lb. weight from the top of the Leaning Tower, and both weights reached the pavement together. By this and other ingenious experiments he laid a firm foundation for the science of mechanics, and he discovered the laws of motion which Newton afterwards formulated. He heard that an instrument had been invented in Holland which seemed to bring distant objects nearer, and, having himself a considerable knowledge of optics, it was not long before he made himself a little telescope. He fixed two spectacle glasses, one for long and one for short sight, in a little old organ-pipe, and thus made for himself a telescope which magnified three times. Before long he had made another which magnified thirty times, and, turning it towards the heavenly bodies, he discovered dark moving spots upon the Sun, mountains and valleys on the Moon, and four small satellites revolving round Jupiter. He also perceived that Venus showed "**phases**"—that is to say, she changed her apparent shape just as the Moon does—and he found the Milky Way to be composed of an immense number of small stars. These discoveries were made in the years 1609-11.

A telescope consists in principle of two parts—an **object-glass**, to form an image of the distant object, and an **eye-piece**, to magnify it. The rays of light from the heavenly body fall on the object-glass, and are so bent out of their course by it as to be brought together in a point called the focus. The "light-gathering power" of the telescope, therefore, depends upon the size of the object-glass, and is proportional to its area. But the size of the image depends upon the focal length of the telescope, *i.e.* upon the distance that the focus is from the object-glass. Thus a small disc, an inch in diameter—such as a halfpenny—will exactly cover the full Moon if held up nine feet away from the eye; and necessarily the image of the full Moon made by an object-glass of nine-feet focus will be an inch in diameter. The eye-piece is a magnifying-glass or small microscope applied to this image, and by it the image can be magnified to any desired amount which the quality of the object-glass and the steadiness of the atmosphere may permit.

This little image of the Moon, planet, or group of stars lent itself to measurement. A young English gentleman, GASCOIGNE, who afterwards fell at the Battle of Marston Moor, devised the "micrometer" for this purpose. The micrometer usually has two frames, each carrying one or more very thin threads—usually spider's threads—and the frames can be moved by very fine screws, the number of turns or parts of a turn of each screw being read off on suitable scales. By placing one thread on the image of one star, and the other on the image of another, the apparent separation of the two can be readily and precisely measured.

Within the last thirty years photography has immensely increased the ease with which astronomical measurements can be made. The sensitive photographic plate is placed in the focus of the telescope, and the light of Sun, Moon, or stars, according to the object to which the telescope is directed, makes a permanent impression on the plate. Thus a picture is obtained, which can be examined and measured in detail at any convenient time afterwards; a portion of the heavens is, as it were, brought actually down to the astronomer's study.

It was long before this great advance was effected. The first telescopes were very imperfect, for the rays of different colour proceeding from any planet or star came to different foci, so that the image was coloured, diffused, and ill-defined. The first method by which this difficulty was dealt with was by making telescopes of enormously long focal length; 80, 100, or 150 feet were not uncommon, but these were at once cumbersome and unsteady. Sir Isaac Newton therefore discarded the use of object-glasses, and used curved mirrors in order to form the image in the focus, and succeeded in making two telescopes on this principle of reflection. Others followed in the same direction, and a century later Sir WILLIAM HERSCHEL was most skilful and successful in making "**reflectors**," his largest being 40 feet in focal length, and thus giving an image of the Moon in its focus of nearly 4-½ inches diameter.

But in 1729 CHESTER MOOR HALL found that by combining two suitable lenses together in the object-glass he could get over most of the colour difficulty, and in 1758 the optician DOLLOND began to make object-glasses that were almost free from the colour defect. From that time onward the manufacture of "**refractors**," as object-glass telescopes are called, has improved; the glass has been made more transparent and more perfect in quality, and larger in size, and the figure of the lens improved. The largest refractor now in use is that of the Yerkes Observatory, Wisconsin, U.S.A., and is 40 inches in aperture, with a focal length of 65 feet, so that the image of the Moon in its focus has a diameter of more than 7 inches. At present this seems to mark the limit of size for refractors, and the difficulty of getting good enough glass for so large a lens is very great indeed. Reflectors have therefore come again into favour, as mirrors can be made larger than any object-glass. Thus Lord Rosse's great telescope was 6 feet in diameter; and the most powerful telescope now in action is the great 5-foot mirror of the Mt. Wilson Observatory, California, with a focal length, as sometimes used, of 150 feet. Thus its light-gathering power is about 60,000 times that of the unaided eye, and the full Moon in its focus is 17 inches in diameter; such is the enormous increase to man's power of sight, and consequently to his power of learning about the heavenly bodies, which the development of the telescope has afforded to him.

The measurement of time was the first purpose for which men watched the heavenly bodies; a second purpose was the measurement of the size of the Earth. If at one place a star was observed to pass exactly overhead, and if at another, due south of it, the same star was observed to pass the meridian one degree north of the zenith, then by measuring the distance between the two places the circumference of the whole Earth would be known, for it would be 360 times that amount. In this way the size of the Earth was roughly ascertained 2000 years before the invention of the telescope. But with the telescope measures of much greater precision could be made, and hence far more difficult problems could be attacked.

One great practical problem was that of finding out the position of a ship when out of sight of land. The ancient Phoenician and Greek navigators had mostly confined themselves to coasting voyages along the shores of the Mediterranean Sea, and therefore the quick recognition of landmarks was the first requisite for a good sailor. But when, in 1492, Columbus had brought a new continent to light, and long voyages were freely taken across the great oceans, it became an urgent necessity for the navigator to find out his position when he had been out of sight of any landmark for weeks.

This necessity was especially felt by the nations of Western Europe, the countries facing the Atlantic with the New World on its far-distant other shore. Spain, France, England, and Holland, all were eager competitors for a grasp on the new lands, and therefore were earnest in seeking a solution of the problem of navigation.

The latitude of the ship could be found out by observing the height of the Sun at noon, or of the Pole Star at night, or in several other ways. But the longitude was more difficult. As the Earth turns on its axis, different portions of its surface are brought in succession under the Sun, and if we take the moment when the Sun is on the meridian of any place as its noon, as twelve o'clock for that place, then the difference of longitude between any two places is essentially the difference in their local times.

It was possible for the sailor to find out when it was local noon for him, but how could he possibly find out what time it was at that moment at the port from which he had sailed, perhaps several weeks before?

The Moon and stars supplied eventually the means for giving this information. For the Moon moves amongst the stars, as the hand of a clock moves amongst the figures of a dial, and it became possible at length to predict for long in advance exactly where amongst the stars the Moon would be, for any given time, of any selected place.

When this method was first suggested, however, neither the motion of the Moon nor the places of the principal stars were known with sufficient accuracy, and it was to remedy this defect, and put navigation upon a sound basis, that CHARLES II. founded Greenwich Observatory in the year 1675, and appointed FLAMSTEED the first Astronomer Royal. In the year 1767 MASKELYNE, the fifth Astronomer Royal, brought out the first volume of the *Nautical Almanac,* in which the positions of the Moon relative to certain stars were given for regular intervals of Greenwich time. Much about the same period the problem was solved in another way by the invention of the chronometer, by JOHN HARRISON, a Yorkshire carpenter. The **chronometer** was a large watch, so constructed that its rate was not greatly altered by heat or cold, so that the navigator had Greenwich time with him wherever he went.

The new method in the hands of CAPTAIN COOK and other great navigators led to a rapid development of navigation and the discovery of Australia and New Zealand, and a number of islands in the Pacific. The building up of the vast oceanic commerce of Great Britain and of her great colonial empire, both in North America and in the Southern Oceans, has arisen out of the work of the Royal Observatory, Greenwich, and has had a real and intimate connection with it.

To observe the motions of the Moon, Sun, and planets, and to determine with the greatest possible precision the places of the stars have been the programme of Greenwich Observatory from its foundation to the present time. Other great national observatories have been Copenhagen, founded in 1637; Paris, in 1667; Berlin, in 1700; St. Petersburg, in 1725, superseded by that of Pulkowa, in 1839; and Washington, in 1842; while not a few of the great universities have also efficient observatories connected with them.

Of the directly practical results of astronomy, the promotion of navigation stands in the first rank. But the science has never been limited to merely utilitarian inquiries, and the problem of measuring celestial distances has followed on inevitably from the measurement of the Earth.

The first distance to be attacked was that of the nearest companion to the Earth, *i.e.* the Moon. It often happens on our own planet that it is required to find the distance of an object beyond our reach. Thus a general on the march may come to a river and need to know exactly how broad it is, that he may prepare the means for bridging it. Such problems are usually solved on the following principle. Let A be the distant object. Then if the direction of A be observed from each of two stations, B and C, and the distance of B from C be measured, it is possible to calculate the distances of A from B and from C. The application of this principle to the measurement of the Moon's distance was made by the establishment of an observatory at the Cape of Good Hope, to co-operate with that of Greenwich. It is, of course, not possible to see Greenwich Observatory from the Cape, or vice versa, but the stars, being at an almost infinite distance, lie in the same direction from both observatories. What is required then is to measure the apparent distance of the Moon from the same stars as seen from Greenwich and as seen from the Cape, and, the distance apart of the two observatories being known, the distance of the Moon can be calculated.

This was a comparatively easy problem. The next step in celestial measurement was far harder; it was to find the distance of the Sun. The Sun is 400 times as far off as the Moon, and therefore it seems to be practically in the same direction as seen from each of the two observatories, and, being so bright, stars cannot be seen near it in the telescope. But by carefully watching the apparent movements of the planets their *relative* distances from the Sun can be ascertained, and were known long before it was thought possible that we should ever know their real distances. Thus Venus never appears to travel more than 47° 15′ from the Sun. This means that her distance from the Sun is a little more than seven-tenths of that of the Earth. If, therefore, the distance of one planet from the Sun can be measured, or the distance of one planet from the Earth, the actual distances of all the planets will follow. We know the proportions of the parts of the solar system, and, if we can fix the scale of one of the parts, we fix the scale of all.

It has been found possible to determine the distance of Mars, of several of the "minor planets," and especially of Eros, a very small minor planet that sometimes comes within 13,000,000 miles of the Earth, or seven times nearer to us than is the Sun.

From the measures of Eros, we have learned that the Sun is separated from us by very nearly 93,000,000 miles—an unimaginable distance. Perhaps the nearest way of getting some conception of this vast interval is by remembering that there are only 31,556,926 seconds of time in a year. If, therefore, an express train, travelling 60 miles an hour—a mile a minute—set out for the Sun, and travelled day and night without cease, it would take more than 180 years to accomplish the journey.

But this astronomical measure has led on to one more daring still. The earth is on one side of the Sun in January, on the other in July. At these two dates, therefore, we are occupying stations 186,000,000 miles apart, and can ascertain the apparent difference in direction of the stars as viewed from the two points But the astonishing result is that this enormous change in the position of the Earth makes not the slightest observable difference in the position of most of the stars. A few, a very few, do show a very slight difference. The nearest star to us is about 280,000 times as far from us as the Sun; this is Alpha Centauri, the brightest star in the constellation of the Centaur and the third brightest star in the sky. Sirius, the brightest star, is twice this distance. Some forty or fifty stars have had their distances roughly determined; but the stars in general far transcend all our attempts to plumb their distances. But, from certain indirect hints, it is generally supposed that the mass of stars in the Milky Way are something like 300,000,000 times as far from us as we are from our Sun.

Thus far, then, astronomy has led us in the direction or measurement. It has enabled us to measure the size of the Earth upon which we live, and to find out the position of a ship in the midst of the trackless ocean. It has also enabled us to cast a sounding-line into space, to show how remote and solitary the earth moves through the void, and to what unimaginable lengths the great stellar universe, of which it forms a secluded atom, stretches out towards infinity.

# CHAPTER V
# THE MEMBERS OF THE SOLAR SYSTEM

Astronomical measurement has not only given us the distances of the various planets from the Sun; it has also furnished us, as in the annexed table, with their real diameters, and, as a consequence of the law of gravitation, with their densities and weights, and the force of gravity at their surfaces. And these numerical details are of the first importance in directing us as to the inferences that we ought to draw as to their present physical conditions.

The theory of Copernicus deprived the Earth of its special position as the immovable centre of the universe, but raised it to the rank of a planet. It is therefore a heavenly body, yet needs no telescope to bring it within our ken; bad weather does not hide it from us, but rather shows it to us under new conditions. We find it to be a globe of land and water, covered by an atmosphere in which float changing clouds; we have mapped it, and we find that the land and water are always there, but their relations are not quite fixed; there is give and take between them. We have found of what elements the land and water consist, and how these elements combine with each other or dissociate. In a word, the Earth is the heavenly body that we know the best, and with it we must compare and contrast all the others.

Before the invention of the telescope there were but two other heavenly bodies—the Sun and the Moon—that appeared as orbs showing visible discs, and even in their cases nothing could be satisfactorily made out as to their conditions. Now each of the five planets known to the ancients reveals to us in the telescope a measurable disc, and we can detect significant details on their surfaces.

THE MOON is the one object in the heavens which does not disappoint a novice when he first sees it in the telescope. Every detail is hard, clear-cut, and sharp; it is manifest that we are looking at a globe, a very rough globe, with hills and mountains, plains and valleys, the whole in such distinct relief that it seems as if it might be touched. No clouds ever conceal its details, no mist ever softens its outlines; there are no half-lights, its shadows are dead black, its high lights are molten silver. Certain changes of illumination go on with the advancing age of the Moon, as the crescent broadens out to the half, the half to the full, and the full, in its turn, wanes away; but the lunar day is nearly thirty times as long as that of the Earth, and these changes proceed but slowly.

The full Moon, as seen by the naked eye, shows several vague dark spots, which most people agree to fancy as like the eyes, nose, and mouth of a broad, sorrowful face. The ordinary astronomical telescope inverts the image, so the "eyes" of the Moon are seen in the lower part of the field of the telescope as a series of dusky plains stretching right across the disc. But in the upper part, near the left-hand corner of the underlip, there is a bright, round spot, from which a number of bright streaks radiate—suggesting a peeled orange with its stalk, and the lines marking the sections radiating from it. This bright spot has been called after the great

| Class. | Name. | Mean distance from Sun. Earth's distance = 1. | In millions of miles. | Period of revolution. In years. | Velocity in orbit. Miles per sec. | Eccentricity. |
|---|---|---|---|---|---|---|
| Terrestrial Planets | Mercury | 0.387 | 36.0 | 0.24 | 29.7 | 0.2056 |
| | Venus | 0.723 | 67.2 | 0.62 | 21.9 | 0.0068 |
| | Earth | 1.000 | 92.9 | 1.00 | 18.5 | 0.0168 |
| | Mars | 1.524 | 141.5 | 1.88 | 15.0 | 0.0933 |
| Minor Planets | Eros | 1.458 | 135.5 | 1.76 | 15.5 | 0.2228 |
| | Ceres | 2.767 | 257.1 | 4.60 | 11.1 | 0.0763 |
| | Achilles | 5.253 | 488.0 | 12.04 | 8.1 | 0.0509 |
| Major Planets | Jupiter | 5.203 | 483.3 | 11.86 | 8.1 | 0.0483 |
| | Saturn | 9.539 | 886.6 | 29.46 | 6.0 | 0.0561 |
| | Uranus | 19.183 | 1781.9 | 84.02 | 4.2 | 0.0463 |
| | Neptune | 30.055 | 2791.6 | 164.78 | 3.4 | 0.0090 |

| Name. | Symbol. | Mean diameter. In miles. | [Earth]=1. | Surface. [Earth]=1. | Volume. [Earth]=1. | Mass. [Earth]=1. |
|---|---|---|---|---|---|---|
| Sun | [Sun] | 866400 | 109.422 | 11973. | 1310130. | 332000. |
| Moon | [Moon] | 2163 | 0.273 | 0.075 | 0.02 | 0.012 |
| Mercury | [Mercury] | 3030 | 0.383 | 0.147 | 0.06 | 0.048 |
| Venus | [Venus] | 7700 | 0.972 | 0.945 | 0.92 | 0.820 |
| Earth | [Earth] | 7918 | 1.000 | 1.000 | 1.00 | 1.000 |
| Mars | [Mars] | 4230 | 0.534 | 0.285 | 0.15 | 0.107 |
| Jupiter | [Jupiter] | 86500 | 10.924 | 119.3 | 1304. | 317.7 |
| Saturn | [Saturn] | 73000 | 9.219 | 85.0 | 783. | 94.8 |
| Uranus | [Uranus] | 31900 | 4.029 | 16.2 | 65. | 14.6 |
| Neptune | [Neptune] | 34800 | 4.395 | 19.3 | 85. | 17.0 |

| Name. | Density. [Earth] =1. | Density. Water =1. | Gravity. [Earth] =1. | Gravity. Fall in feet per sec. | Light and heat received from Sun. [Earth]=1. | Time of rotation on axis. | Albedo; *i.e.* reflecting power. |
|---|---|---|---|---|---|---|---|
| | | | | | | d. h. m. | |
| Sun | 0.25 | 1.39 | 27.65 | 444.60 | ... | 25 4 48 ± | ... |
| Moon | 0.61 | 3.39 | 0.17 | 2.73 | 1.00 | 27 7 43 | 0.17 |
| | | | | | | d. h. m. s. | |
| Mercury | 0.85 | 4.72 | 0.43 | 6.91 | 6.67 | 88 (?) | 0.14 |
| Venus | 0.89 | 4.94 | 0.82 | 13.19 | 1.91 | 23 21 23 (?) | 0.76 |
| Earth | 1.00 | 5.55 | 1.00 | 16.08 | 1.00 | 23 56 4 | 0.50 (?) |
| Mars | 0.71 | 3.92 | 0.38 | 6.11 | 0.43 | 24 37 23 | 0.22 |
| | | | | | | h. m. | |
| Jupiter | 0.24 | 1.32 | 2.65 | 42.61 | 0.037 | 9 55 ± | 0.62 |
| Saturn | 0.13 | 0.72 | 1.18 | 18.97 | 0.011 | 10 14 ± | 0.72 |
| Uranus | 0.22 | 1.22 | 0.90 | 14.47 | 0.003 | 9 30 (?) | 0.60 |
| Neptune | 0.20 | 1.11 | 0.89 | 14.31 | 0.001 | (?) | 0.52 |

Danish astronomer, "Tycho," and is one of the most conspicuous objects of the full Moon.

The contrasts of the Moon are much more pronounced when she is only partly lit up. Then the mountains and valleys stand out in the strongest relief, and it becomes clear that the general type of formation on the Moon is that of rings—rings of every conceivable size, from the smallest point that the telescope can detect up to some of the great dusky plains themselves, hundreds of miles in diameter. These rings are so numerous that Galileo described the Moon as looking as full of "eyes" as a peacock's tail.

The "right eye" of the moonface, as we see it in the sky, is formed by a vast dusky plain, nearly as large as France and Germany put together, to which has been given the name of the "Sea of Rains" (*Mare Imbrium*), and just below this (as seen in the telescope) is one of the most perfect and beautiful of all the lunar rings—a great ring-plain, 56 miles in diameter, called after the thinker who revolutionised men's ideas of the solar system, "Copernicus." "Copernicus," like "Tycho," is the centre of a set of bright streaks; and a neighbouring but smaller ring, bearing the great name of "Kepler," stands in a like relation to another set.

The most elevated region of the Moon is immediately in the neighbourhood of the great "stalk of the orange," "Tycho." Here the rings are crowded together as closely as they can be packed; more closely in many places, for they intrude upon and overlap each other in the most intricate manner. A long chain of fine rings stretches from this disturbed region nearly to the centre of the disc, where the great Alexandrian astronomer is commemorated by a vast walled plain, considerably larger than the whole of Wales, and known as "Ptolemæus."

The distinctness of the lunar features shows at once that the Moon is in an altogether different condition from that of the Earth. Here the sky is continually being hidden by cloud, and hence the details of the surface of the Earth as viewed from any other planet must often be invisible, and even when actual cloud is absent there is a more permanent veil of dust, which must greatly soften and confuse terrestrial outlines. The clearness, therefore, with which we perceive the lunar formations proves that there is little or no atmosphere there. Nor is there any sign upon it of water, either as seas or lakes or running streams.

Yet the Moon shows clearly that in the past it has gone through great and violent changes. The gradation is so complete from the little craterlets, which resemble closely, in form and size, volcanic craters on the Earth, up to the great ring-plains, like "Copernicus" or "Tycho," or formations larger still, that it seems natural to infer not only that the smaller craters were formed by volcanic eruption, like the similar objects with which we are acquainted on our own Earth, but that the others, despite their greater sizes, had a like origin. In consequence of the feebler force of gravity on the Moon, the same explosive force there would carry the material of an eruption much further than on the Earth.

The darker, low-lying districts of the Moon give token of changes of a different order. It is manifest that the material of which the floors of these plains is composed has invaded, broken down, and almost submerged many of the ring-formations. Sometimes half of a ring has been washed away; sometimes just the outline of a ring can still be traced upon the floor of the sea; sometimes only a slight breach has been made in the wall. So it is clear that the Moon was once richer in the great crater-like formations than it is to-day, but a lava-flood has overflowed at least one-third of its area. More recent still are the bright streaks, or rays, which radiate in all directions from "Tycho," and from some of the other ring-plains.

It is evident from these different types of structure on the Moon, and from the relations which they bear to each other, that the lunar surface has passed through several successive stages, and that its changes tended, on the whole, to diminish in violence as time went on; the minute crater pits with which the surface is stippled having been probably the last to form.

But the 300 years during which the Moon has been watched with the telescope have afforded no trace of any continuance of these changes. She has had a stormy and fiery past; but nothing like the events of those bygone ages disturbs her serenity to-day.

And yet we must believe that change does take place on the Moon even now, because during the 354 hours of its long day the Sun beats down with full force on the unprotected surface, and during the equally long night that surface is exposed to the cold of outer space. Every part of the surface must be exposed in turn to an extreme range of temperature, and must be cracked, torn, and riven by alternate expansion and contraction. Apart from this slow, wearing process, and a very few rather doubtful cases in which a minute alteration of some surface detail has been suspected, our sister planet, the Moon, shows herself as changeless and inert, without any appreciable trace of air or water or any sign of life—a dead world, with all its changes and activities in the past.

MARS, after the Moon, is the planet whose surface we can study to best advantage. Its orbit lies outside that of the Earth, so that when it is nearest to us it turns the same side to both the Sun and Earth, and we see it fully illuminated. Mercury and Venus, on the contrary, when nearest us are between us and the Sun, and turn their dark sides to us. When fully illuminated they are at their greatest distance, and appear very small, and, being near the Sun, are observed with difficulty. These three are intermediate in size between the Moon and the Earth.

In early telescopic days it was seen that Mars was an orange-coloured globe with certain dusky markings upon it, and that these markings slowly changed their place—that, in short, it was a world rotating upon its axis, and in a period not very different from that of the Earth. The rotation period of Mars has indeed been fixed to the one-hundredth part of a second of time; it is 24 h. 37 m. 22.67 s. And this has been possible because some of the dusky spots observed in the seventeenth century can be identified now in the twentieth. Some of the markings on Mars, like our own continents and seas, and like the craters on the Moon, are permanent features; and many charts of the planet have been constructed.

Other markings are variable. Since the planet rotates on its axis, the positions of its poles and equator are known, its equator being inclined to its orbit at an angle of 24° 50′, while the angle in the case of the Earth is 23° 27′. The times when its seasons begin and end are therefore known; and it is found that the spring of its northern hemisphere lasts 199 of our days, the summer 183, the autumn 147, and the winter 158. Round the pole in winter a broad white cap forms, which begins to shrink as spring comes on, and may entirely disappear in summer. No corresponding changes have been observed on the Moon, but it is easy to find an analogy to them on the Earth. Round both our poles a great cap of ice and snow is spread—a cap which increases in size as winter comes on, and diminishes with the advance of summer—and it seems a reasonable inference to suppose that the white polar caps of Mars are, like our own, composed of ice and snow.

From time to time indications have been observed of the presence on Mars of a certain amount of cloud. Familiar dark markings have, for a short time, been interrupted, or been entirely hidden, by white bands, and have recovered their ordinary appearance later. With rotation on its axis and succession of seasons, with atmosphere and cloud, with land and water, with ice and snow, Mars would seem to be a world very similar to our own.

This was the general opinion up to the year 1877, when SCHIAPARELLI announced that he had discovered a number of very narrow, straight, dark lines on the planet—lines to which he gave the name of "canali"—that is, "channels." This word was unfortunately rendered into English by the word "**canals**," and, as a canal means a waterway artificially made, this mistranslation gave the idea that Mars was inhabited by intelligent beings, who had dug out the surface of the planet into a network of canals of stupendous length and breadth.

The chief advocate of this theory is LOWELL, an American observer, who has given very great attention to the study of the planet during the last seventeen years. His argument is that the straight lines, the canals, which he sees on the planet, and the round dots, the "**oases**," which he finds at their intersections, form a system so obviously *un*natural, that it must be the work of design—of intelligent beings. The canals are to him absolutely regular and straight, like lines drawn with ruler and pen-and-ink, and the oases are exactly round. But, on the one hand, the best observers, armed with the most powerful telescopes, have often been able to perceive that markings were really full of irregular detail, which Lowell has represented as mere hard straight lines and circular dots, and, on the other hand, the straight line and the round dot are the two geometric forms which all very minute objects must approach in appearance. That we cannot see irregularities in very small and distant objects is no proof at all that irregularities do not exist in them, and it has often happened that a marking which appeared a typical "canal" when Mars was at a great distance lost that appearance when the planet was nearer.

Astronomers, therefore, are almost unanimous that there is no reason for supposing that any of the details that we see on the surface of Mars are artificial in their origin. And indeed the numerical facts that we know about the planet render it almost impossible that there should be any life upon it.

If we turn to the table, we see that in size, volume, density, and force of gravity at its surface, Mars lies between the Moon and the Earth, but is nearer the Moon. This has an important bearing as to the question of the planet's atmosphere. On the Earth we pass through half the atmosphere by ascending a mountain that is three and a third miles in height; on Mars we should have to ascend nearly nine miles. If the atmospheric pressure at the surface of Mars were as great as it is at the surface of the Earth, his atmosphere would be far deeper than ours and would veil the planet more effectively. But we see the surface of Mars with remarkable distinctness, almost as clearly, when its greater distance is allowed for, as we see the Moon. It is therefore accepted that the atmospheric pressure at the surface of Mars must be very slight, probably much less than at the top of our very highest mountains, where there is eternal snow, and life is completely absent.

But Mars compares badly with the Earth in another respect. It receives less light and heat from the Sun in the proportion of three to seven. This we may express by saying that Mars, on the whole, is almost as much worse off than the Earth as a point on the Arctic Circle is worse off than a point on the Equator. The mean temperature of the Earth is taken as about 60° of the Fahrenheit thermometer (say, 15° Cent.); the mean temperature of Mars must certainly be considerably below freezing-point, probably near 0° F. Here on our Earth the boiling-point of water is 212°, and, since the mean temperature is 60° and water freezes at 32°, it is normally in the liquid state. On Mars it must normally be in the solid state—ice, snow, or frost, or the like. But with so rare an atmosphere water will boil at a low temperature, and it is not impossible that under the direct rays of the Sun—that is to say, at midday of the torrid zone of Mars—ice may not only melt, but water boil by day, condensing and freezing again during the night. NEWCOMB, the foremost astronomer of his day, concluded "that during the night of Mars, even in the equatorial regions, the surface of the planet probably falls to a lower temperature than any we ever experienced on our globe. If any water exists, it must not only be frozen, but the temperature of the ice must be far below the freezing point.... The most careful calculation shows that if there are any considerable bodies of water on our neighbouring planet, they exist in the form of ice, and can never be liquid to a depth of more than one or two inches, and that only within the torrid zone and during a few hours each day." With regard to the snow caps of Mars, Newcomb thought it not possible that any considerable fall of snow could ever take place. He regarded the white caps as simply due to a thin deposit of hoar frost, and it cannot be deemed wonderful that such should gradually disappear, when it is remembered that each of the two poles of Mars is in turn presented to the Sun for more than 300 consecutive days. Newcomb's conclusion was: "Thus we have a kind of Martian meteorological changes, very slight indeed, and seemingly very different from those of our Earth, but yet following similar lines on their small scale. For snowfall substitute frostfall; instead of (the barometer reading) feet or inches say fractions of a millimetre, and instead of storms or wind substitute little motions of an air thinner than that on the top of the Himalayas, and we shall have a general description of Martian meteorology."

We conclude, then, that Mars is not so inert a world as the Moon, but, though some slight changes of climate or weather take place upon it, it is quite unfitted for the nourishment and development of the different forms of organic life.

Of MERCURY we know very little. It is smaller than Mars but larger than the Moon, but it differs from them both in that it is much nearer the Sun, and receives, therefore, many times the light and heat, surface for surface. We should expect, therefore, that water on Mercury would exist in the gaseous state instead of in the solid state as on Mars. The little planet reflects the sunlight only feebly, and shows no evidence of cloud. A few markings have been made out on its surface, and the best observers agree that it appears to turn the same face always to the Sun. This would imply that the one hemisphere is in perpetual darkness and cold, the other, exposed to an unimaginable fiery heat.

VENUS is nearly of the same size as the Earth, and the conditions as to the arrangement of its atmosphere, the force of gravity at its surface, must be nearly the same as on our own world. But we know almost nothing of the details of its surface; the planet is very bright, reflecting fully seven-tenths of the sunlight that falls upon it. It would seem that, in general, we see nothing of the actual details of the planet, but only the upper surface of a very cloudy atmosphere. Owing to the fact that Venus shows no fixed definite marking that we can watch, it is still a matter of controversy as to the time in which it rotates upon its axis. Schiaparelli and some other observers consider that it rotates in the same time as it revolves round the Sun. Others believe that it rotates in a little less than twenty-four hours. If this be so, and there is any body in the solar system other than the Earth, which is adapted to be the home of life, then the planet Venus is that one.

THE SUN, like the Moon, presents a visible surface to the naked eye, but one that shows no details. In the telescope the contrast between it and the Moon is very great, and still greater is the contrast which is brought out by the measurements of its size, volume, and weight. But the really significant difference is that the Sun is a body giving out light and heat, not merely reflecting them. Without doubt this last difference is connected most closely with the difference in size. The Moon is cold, dead, unchanging, because it is a small world; the Sun is bright, fervent, and undergoes the most violent change, because it is an exceedingly large world.

The two bodies—the Sun and Moon—appear to the eye as being about the same size, but since the Sun is 400 times as far off as the Moon it must be 400 times the diameter. That means that it has 400 times 400, or 160,000 times the surface and 400 times 400 times 400, or 64,000,000 times the volume. The Sun and Moon, therefore, stand at the very extremes of the scale.

The heat of the Sun is so great that there is some difficulty in observing it in the telescope. It is not sufficient to use a dark glass in order to protect the eye, unless the telescope be quite a small one. Some means have to be employed to get rid of the greater part of the heat and light. The simplest method of observing is to fix a screen behind the eyepiece of a telescope and let the image of the Sun be projected upon the screen, or the sensitive plate may be substituted for the screen, and a photograph obtained, which can be examined at leisure afterwards.

As generally seen, the surface of the Sun appears to be mottled all over by a fine irregular stippling. This stippling, though everywhere present, is not very strongly marked, and a first hasty glance might overlook it. From time to time, however, dark spots are seen, of ever-changing form and size. By watching these, Galileo proved that the Sun rotated on its axis in a little more than twenty-five days, and in the nineteenth century SCHWABE proved that the sunspots were not equally large and numerous at all times, but that there was a kind of cycle of a little more than eleven years in average length. At one time the Sun would be free from spots; then a few small ones would appear; these would gradually become larger and more numerous; then a decline would follow, and another spotless period would succeed about eleven years after the first. As a rule, the increase in the spots takes place more quickly than the decline.

Most of the spot-groups last only a very few days, but about one group in four lasts long enough to be brought round by the rotation of the Sun a second time; in other words, it continues for about a month. In a very few cases spots have endured for half a year.

An ordinary form for a group of spots is a long stream drawn out parallel to the Sun's equator, the leading spot being the largest and best defined. It is followed by a number of very small irregular and ill-developed spots, and the train is brought up by a large spot, sometimes even larger than the leader, but by no means so regular in form or so well defined. The leading spot for a short time moves forward much faster than its followers, at a speed of about 8000 miles per day. The small middle spots then gradually die out, or rather seem to be overflowed by the bright material of the solar surface, the "**photosphere**," as it is called; the spot in the rear breaks up a little later, and the leader, which is now almost circular, is left alone, and may last in this condition for some weeks. Finally, it slowly contracts or breaks up, and the disturbance comes to an end. This is the course of development of many long-lived spot-groups, but all do not conform to the same type. The very largest spots are indeed usually quite different in their appearance and history.

In size, sunspots vary from the smallest dot that can be discovered in the telescope up to huge rents with areas that are to be counted by thousands of millions of square miles; the great group of February 1905 had an area of 4,000,000,000 square miles, a thousand times the area of Europe.

Closely associated with the *maculæ*, as the spots were called by the first observers, are the "**faculæ**"—long, branching lines of bright white light, bright as seen even against the dazzling background of the Sun itself, and looking like the long lines of foam of an incoming tide. These are often associated with the spots; the spots are formed between their ridges, and after a spot-group has disappeared the broken waves of faculæ will sometimes persist in the same region for quite a long time.

The faculæ clearly rise above the ordinary solar surface; the spots as clearly are depressed a little below it; because from time to time we see the bright material of the surface pour over spots, across them, and sometimes into them. But there is no reason to believe that the spots are deep, in proportion either to the Sun itself or even to their own extent.

Sunspots are not seen in all regions of the Sun. It is very seldom that they are noted in a higher solar latitude than 40°, the great majority of spots lying in the two zones between 5° and 25° latitude on either side of the equator. Faculæ, on the other hand, though most frequent in the spot zones, are observed much nearer the two poles.

It is very hard to find analogies on our Earth for sunspots and for their peculiarities of behaviour. Some of the earlier astronomers thought they were like terrestrial volcanoes, or rather like the eruptions from them. But if there were a solid nucleus to the Sun, and the spots were eruptions from definite areas of the nucleus, they would all give the same period of rotation. But sunspots move about freely on the solar surface, and the different zones of that surface rotate in different times, the region of the equator rotating the most quickly. This alone is enough to show that the Sun is essentially not a solid body. Yet far down below the photosphere something approaching to a definite structure must already be forming. For there is a well-marked progression in the zones of sunspots during the eleven-year cycle. At a time when spots are few and small, known as **the sunspot minimum**, they begin to be seen in fairly high latitudes. As they get more numerous, and many of them larger, they frequent the medium zones. When the Sun is at its greatest activity, known as **the sunspot maximum**, they are found from the highest zone right down to the equator. Then the decline sets in, but it sets in first in the highest zones, and when the time of minimum has come again the spots are close to the equator. Before these have all died away, a few small spots, the heralds of a new cycle of activity, begin to appear in high latitudes.

This law, called after SPÖRER, its discoverer, indicates that the origin and source of sunspot activity lie within the Sun. At one time it was thought that sunspots were due to some action of Jupiter—for Jupiter moves round the Sun in 11.8 years, a period not very different from the sunspot cycle—or to some meteoric stream. But Spörer's Law could not be imposed by some influence from without. Still sunspots, once formed, may be influenced by the Earth, and perhaps by other planets also, for MRS. WALTER MAUNDER has shown that the numbers and areas of spots tend to be smaller on the western half of the disc, as seen from the Earth, than on the eastern, while considerably more groups come into view at the east edge of the Sun than pass out of view at the west edge, so that it would appear as if the Earth had a damping effect upon the spots exposed to it.

But the Sun is far greater than it ordinarily appears to us. Twice every year, and sometimes oftener, the Moon, when new, comes between the Earth and the Sun, and we have an **Eclipse of the Sun**, the dark body of the Moon hiding part, or all, of the greater light. The Sun and Moon are so nearly of the same apparent size that an eclipse of the Sun is total only for a very narrow belt of the Earth's surface, and, as the Moon moves more quickly than the Sun, the eclipse only remains total for a very short time—seven minutes at the outside, more usually only two or three. North or south of that belt the Moon is projected, so as to leave uncovered a part of the Sun north or south of the Moon. A total eclipse, therefore, is rare at any particular place, and if a man were able to put himself in the best possible position on each occasion, it would cost him thirty years to secure an hour's accumulated duration.

Eclipses of the Moon are visible over half the world at one time, for there is a real loss to the Moon of her light. Her eclipses are brought about when, in her orbit, she passes behind the Earth, and the Earth, being between the Sun and the Moon, cuts off from the latter most of the light falling upon her;

not quite all; a small portion reaches her after passing through the thickest part of the Earth's atmosphere, so that the Moon in an eclipse looks a deep copper colour, much as she does when rising on a foggy evening.

Total eclipses of the Sun have well repaid all the efforts made to observe them. It is a wonderful sight to watch the blackness of darkness slowly creeping over the very fountain of light until it is wholly and entirely hidden; to watch the colours fade away from the landscape and a deathlike, leaden hue pervade all nature, and then to see a silvery, star-like halo, flecked with bright little rose-coloured flames, flash out round the black disc that has taken the place of the Sun.

These rose-coloured flames are the solar "**prominences**," and the halo is the "**corona**," and it is to watch these that astronomers have made so many expeditions hither and thither during the last seventy years. The "prominences," or red flames, can be observed, without an eclipse, by means of the spectroscope, but, as the work of the spectroscope is to form the subject of another volume of this series, it is sufficient to add here that the prominences are composed of various glowing gases, chiefly of hydrogen, calcium, and helium.

These and other gases form a shell round the Sun, about 3000 miles in depth, to which the name "**chromosphere**" has been given. It is out of the chromosphere that the prominences arise as vast irregular jets and clouds. Ordinarily they do not exceed 40 or 50 thousand miles in height, but occasionally they extend for 200 or even 300 thousand miles from the Sun. Their changes are as remarkable as their dimensions; huge jets of 50 or 100 thousand miles have been seen to form, rise, and disappear within an hour or less, and movements have been chronicled of 200 or 300 miles in a single second of time.

Prominences are largest and most frequent when sunspots and faculæ are most frequent, and fewest when those are fewest. The corona, too, varies with the sunspots. At the time of maximum the corona sends forth rays and streamers in all directions, and looks like the conventional figure of a star on a gigantic scale. At minimum the corona is simpler in form, and shows two great wings, east and west, in the direction of the Sun's equator, and round both of his poles a number of small, beautiful jets like a crest of feathers.

Some of the streamers or wings of the corona have been traced to an enormous distance from the Sun. Mrs. Walter Maunder photographed one ray of the corona of 1898 to a distance of 6 millions of miles. LANGLEY, in the clear air of Pike's Peak, traced the wings of the corona of 1878 with the naked eye to nearly double this distance.

But the rapid changes of sunspots and the violence of some of the prominence eruptions are but feeble indications of the most wonderful fact concerning the Sun, *i.e.* the enormous amount of light and heat which it is continually giving off. Here we can only put together figures which by their vastness escape our understanding. Sunlight is to moonlight as 600,000 is to 1, so that if the entire sky were filled up with full moons, they would not give us a quarter as much light as we derive from the Sun. The intensity of sunlight exceeds by far any artificial light; it is 150 times as bright as the calcium light, and three or four times as bright as the brightest part of the electric arc light. The amount of heat radiated by the Sun has been expressed in a variety of different ways; C. A. YOUNG very graphically by saying that if the Sun were encased in a shell of ice 64 feet deep, its heat would melt the shell in one minute, and that if a bridge of ice could be formed from the Earth to the Sun, 2-½ miles square in section and 93 millions of miles long, and the entire solar radiation concentrated upon it, in one second the ice would be melted, in seven more dissipated into vapour.

The Earth derives from the Sun not merely light and heat, but, by transformation of these, almost every form of energy manifest upon it; the energy of the growth of plants, the vital energy of animals, are only the energy received from the Sun, changed in its expression.

The question naturally arises, "If the Sun, to which the Earth is indebted for nearly everything, passes through a change in its activity every eleven years or so, how is the Earth affected by it?" It would seem at first sight that the effect should be great and manifest. A sunspot, like that of February 1905, one thousand times as large as Europe, into which worlds as large as our Earth might be poured, like peas into a saucer, must mean, one might think, an immense falling off of the solar heat.

Yet it is not so. For even this great sunspot was but small as compared with the Sun as a whole. Had it been dead black, it would have stopped out much less than 1 per cent. of the Sun's heat; and even the darkest sunspot is really very bright. And the more spots there are, the more numerous and brighter are the faculæ; so that we do not know certainly which of the two phases, maximum or minimum, means the greater radiation. If the weather on the Earth answers to the sunspot cycle, the connection is not a simple one; as yet no connection has been proved. Thus two of the worst and coldest summers experienced in England fell the one in 1860, the other in 1879, *i.e.* at maximum and minimum respectively. So, too, the hot summers of 1893 and 1911 were also, the one at maximum and the other at minimum; and ordinary average years have fallen at both the phases just the same.

Yet there is an answer on the part of the Earth to these solar changes. The Earth itself is a kind of magnet, possessing a magnetism of which the intensity and direction is always changing. To watch these changes, very sensitive magnets are set up, and a slight daily to-and-fro swing is noticed in them; this swing is more marked in summer than in winter, but it is also more marked at times of the sunspot maximum than at minimum, showing a dependence upon the solar activity.

Yet more, from time to time the magnetic needle undergoes more or less violent disturbance; in extreme cases the electric telegraph communication has been disturbed all over the world, as on September 25, 1909, when the submarine cables ceased to carry messages for several hours. In most cases when such a "magnetic storm" occurs, there is an unusually large or active spot on the Sun. The writer was able in 1904 to further prove that such "storms" have a marked tendency to recur when the same longitude of the Sun is presented again towards the Earth. Thus in February 1892, when a very large spot was on the Sun, a violent magnetic storm broke out. The spot passed out of sight and the storm ceased, but in the following month, when the spot reached exactly the same apparent place on the Sun's disc, the storm broke out again. Such magnetic disturbances are therefore due to streams of particles driven off from limited areas of the Sun, probably in the same way that the long, straight rays of the corona are driven off. Such streams of particles, shot out into space, do not spread out equally in all directions, like the rays of light and heat, but are limited in direction, and from time to time they overtake the Earth in its orbit, and, striking it, cause a magnetic storm, which is felt all over the Earth at practically the same moment.

JUPITER is, after the Sun, much the largest member of the solar system, and it is a peculiarly beautiful object in the telescope. Even a small instrument shows the little disc striped with many delicately coloured bands or belts, broken by white clouds and dark streaks, like a "windy sky" at sunset. And it changes while being watched, for, though 400,000,000 miles away from us, it rotates so fast upon its axis that its central markings can actually be seen to move.

This rapid rotation, in less than ten hours, is the most significant fact about Jupiter. For different spots have different rotation periods, even in the same latitude, proving that we are looking down not upon any solid surface of Jupiter, but upon its cloud envelope—an envelope swept by its rapid rotation and by its winds into a vast system of parallel currents.

One object on Jupiter, the great "**Red Spot**," has been under observation since 1878, and possibly for 200 years before that. It is a large, oval object fitted in a frame of the same shape. The spot itself has often faded and been lost since 1878, but the frame has remained. The spot is in size and position relative to Jupiter much as Australia is to the Earth, but while Australia moves solidly with the rest of the Earth in the daily rotation, neither gaining on South America nor losing on Africa, the Red Spot on Jupiter sees many other spots and clouds pass it by, and does not even retain the same rate of motion itself from one year to another.

No other marking on Jupiter is so permanent as this. From time to time great round white clouds form in a long series as if shot up from some eruption below, and then drawn into the equatorial current. From time to time the belts themselves change in breadth, in colour, and complexity. Jupiter is emphatically the planet of change.

And such change means energy, especially energy in the form of heat. If Jupiter possessed no heat but that it derived from the Sun, it would be colder than Mars, and therefore an absolutely frozen globe. But these rushing winds and hurrying clouds are evidences of heat and activity—a native heat much above that of our Earth. While Mars is probably nearer to the Moon than to the Earth in its condition, Jupiter has probably more analogies with the Sun.

The one unrivalled distinction of SATURN is its Ring. Nothing like this exists elsewhere in the solar system. Everywhere else we see spherical globes; this is a flat disc, but without its central portion. It surrounds the planet, lying in the plane of its equator, but touches it nowhere, a gap of 7000 miles intervening. It appears to be circular, and is 42,000 miles in breadth.

Yet it is not, as it appears to be, a flat continuous surface. It is in reality made up of an infinite number of tiny satellites, mere dust or pebbles for the most part, but so numerous as to look from our distance like a continuous ring, or rather like three or four concentric rings, for certain divisions have been noticed in it—an inner broad division called after its discoverer, CASSINI, and an outer, fainter, narrower one discovered by ENCKE. The innermost part of the ring is dusky, fainter than the planet or the rest of the ring, and is known as the "crape-ring."

Of Saturn itself we know little; it is further off and fainter than Jupiter, and its details are not so pronounced, but in general they resemble those of Jupiter. The planet rotates quickly—in 10 h. 14 m.—its markings run into parallel belts, and are diversified by spots of the same character as on Jupiter. Saturn is probably possessed of no small amount of native heat.

URANUS and NEPTUNE are much smaller bodies than Jupiter and Saturn, though far larger than the Earth. But their distance from the Earth and Sun makes their discs small and faint, and they show little in the telescope beyond a hint of "belts" like those of Jupiter; so that, as with that planet, the surfaces that they show are almost certainly the upper surfaces of a shell of cloud.

In general, therefore, the rule appears to hold good throughout the solar system that a very large body is intensely hot and in a condition of violent activity and rapid change; that smaller bodies are less hot and less active, until we come down to the smallest, which are cold, inert, and dead. Our own Earth, midway in the series, is itself cold, but is placed at such a distance from the Sun as to receive from it a sufficient but not excessive supply of light and heat, and the changes of the Earth are such as not to prohibit but to nourish and support the growth and development of the various forms of life.

The smallest members of the solar system are known as METEORS. These are often no more than pebbles or particles of dust, moving together in associated orbits round the Sun. They are too small and too scattered to be seen in open space, and become visible to us only when their orbits intersect that of the earth, and the earth actually encounters them. They then rush into our atmosphere at a great speed, and become highly heated and luminous as they compress the air before them; so highly heated that most are vapourised and dissipated, but a few reach the ground. As they are actually moving in parallel paths at the time of one of these encounters, they appear from the effect of perspective to diverge from a point, hence called the "**radiant**." Some showers occur on the same date of every year; thus a radiant in the constellation Lyra is active about April 21, giving us meteors, known as the "Lyrids"; and another in Perseus in August, gives us the "Perseids." Other radiants are active at intervals of several years; the most famous of all meteoric showers, that of the "Leonids," from a radiant in Leo, was active for many centuries every thirty-third year; and another falling in the same month, November, came from a radiant in Andromeda every thirteen years. In these four cases and in some others the meteors have been found to be travelling along the same path as a comet. It is therefore considered that meteoric swarms are due to the gradual break up of comets; indeed the comet of the Andromeda shower, known from one of its observers as "Biela's," was actually seen to divide into two in December 1845, and has not been observed as a comet since 1852, though the showers connected with it, giving us the meteors known as the "Andromedes," have continued to be frequent and rich. Meteors, therefore, are the smallest, most insignificant, of all the celestial bodies; and the shining out of a meteor is the last stage of its history—its death; after death it simply goes to add an infinitesimal trifle to the dust of the earth.

# CHAPTER VI
# THE SYSTEM OF THE STARS

The first step towards our knowledge of the starry heavens was made when the unknown and forgotten astronomers of 2700 B.C. arranged the stars into constellations, for it was the first step towards distinguishing one star from another. When one star began to be known as "the star in the eye of the Bull," and another as "the star in the shoulder of the Giant," the heavens ceased to display an indiscriminate crowd of twinkling lights; each star began to possess individuality.

The next step was taken when Hipparchus made his catalogue of stars (129 B.C.), not only giving its name to each star, but measuring and fixing its place—a catalogue represented to us by that of Claudius Ptolemy (A.D. 137).

The third step was taken when BRADLEY, the third Astronomer Royal, made, at Greenwich, a catalogue of more than 3000 star-places determined with the telescope.

A century later ARGELANDER made the great Bonn Zone catalogue of 330,000 stars, and now a great photographic catalogue and chart of the entire heavens have been arranged between eighteen observatories of different countries. This great chart when complete will probably present 30 millions of stars in position and brightness.

The question naturally arises, "Why so many stars? What conceivable use can be served by catalogues of 30 millions or even of 3000 stars?" And so far as strictly practical purposes are concerned, the answer must be that there is none. Thus MASKELYNE, the fifth Astronomer Royal, restricted his observations to some thirty-six stars, which were all that he needed for his *Nautical Almanac*, and these, with perhaps a few additions, would be sufficient for all purely practical ends.

But there is in man a restless, resistless passion for knowledge—for knowledge for its own sake—that is always compelling him to answer the challenge of the unknown. The secret hid behind the hills, or across the seas, has drawn the explorer in all ages; and the secret hid behind the stars has been a magnet not less powerful. So catalogues of stars have been made, and made again, and enlarged and repeated; instruments of ever-increasing delicacy have been built in order to determine the positions of stars, and observations have been made with ever-increasing care and refinement. It is knowledge for its own sake that is longed for, knowledge that can only be won by infinite patience and care.

The chief instrument used in making a star catalogue is called a transit circle; two great stone pillars are set up, each carrying one end of an axis, and the axis carries a telescope. The telescope can turn round like a wheel, in one direction only; it points due north or due south. A circle carefully divided into degrees and fractions of a degree is attached to the telescope.

In the course of the twenty-four hours every star above the horizon of the observatory must come at least once within the range of this telescope, and at that moment the observer points the telescope to the star, and notes the time by his clock when the star crossed the spider's threads, which are fitted in the focus of his eye-piece. He also notes the angle at which the telescope was inclined to the horizon by reading the divisions of his circle. For by these two—the time when the star passed before the telescope and the angle at which the telescope was inclined—he is able to fix the position of the star.

"But why should catalogues be repeated? When once the position of a star has been observed, why trouble to observe it again? Will not the record serve in perpetuity?"

The answers to these questions have been given by star catalogues themselves, or have come out in the process of making them. The Earth rotates on its axis and revolves round the Sun. But that axis also has a rolling motion of its own, and gives rise to an apparent motion of the stars called **Precession**. Hipparchus discovered this effect while at work on his catalogue, and our knowledge of the amount of Precession enables us to fix the date when the constellations were designed.

Similarly, Bradley discovered two further apparent motions of the stars—**Aberration** and **Nutation**. Of these, the first arises from the fact that the light coming from the stars moves with an inconceivable speed, but does not cross from star to Earth instantly; it takes an appreciable, even a long, time to make the journey. But the Earth is travelling round the Sun, and therefore continually changing its direction of motion, and in consequence there is an apparent change in the direction in which the star is seen. The change is very small, for though the Earth moves 18-½ miles in a second, light travels 10,000 times as fast. Stars therefore are deflected from their true positions by Aberration, by an extreme amount of 20.47" of arc, that being the angle shown by an object that is slightly more distant than 10,000 times its diameter.

The axis of the Earth not only rolls on itself, but it does so with a slight staggering, nodding motion, due to the attractions of the Sun and Moon, known as **Nutation**. And the axis does not remain fixed in the solid substance of the Earth, but moves about irregularly in an area of about 60 feet in diameter. The positions of the north and south poles are therefore not precisely fixed, but move, producing what is known as the **Variation of Latitude**. Then star-places have to be corrected for the effect of our own atmosphere, *i.e.* refraction, and for errors of the instruments by which their places are determined. And when all these have been allowed for, the result stands out that different stars have real movement of their own—their **Proper Motions**.

No stars are really "fixed"; the name "**fixed stars**" is a tradition of a time when observation was too rough to detect that any of the heavenly bodies other than the planets were in motion. But nothing is fixed. The Earth on which we stand has many different motions; the stars are all in headlong flight.

And from this motion of the stars it has been learned that the Sun too moves. When Copernicus overthrew the Ptolemaic theory and showed that the Earth moves round the Sun, it was natural that men should be satisfied to take this as the centre of all things, fixed and immutable. It is not so. Just as a traveller driving through a wood sees the trees in front apparently open out and drift rapidly past him on either hand, and then slowly close together behind him, so Sir WILLIAM HERSCHEL showed that the stars in one part of the heavens appear to be opening out, or slowly moving apart, while in the opposite part there seems to be a slight tendency for them to come together, and in a belt midway between the two the tendency is for a somewhat quicker motion toward the second point. And the explanation is the same in the one case as in the other—the real movement is with the observer. The Sun with all its planets and smaller attendants is rushing onward, onward, towards a point near the borders of the constellations Lyra and Hercules, at the rate of about twelve miles per second.

Part of the Proper Motions of the stars are thus only apparent, being due to the actual motion of the Sun—the "**Sun's Way**," as it is called—but part of the Proper Motions belong to the stars themselves; they are really in motion, and this not in a haphazard, random manner. For recently KAPTEYN and other workers in the same field have brought to light the fact of **Star-Drift**, *i.e.* that many of the stars are travelling in associated companies. This may be illustrated by the seven bright stars that make up the well-known group of the "Plough," or "Charles's Wain," as country people call it. For the two stars of the seven that are furthest apart in the sky are moving together in one direction, and the other five in another.

Another result of the close study of the heavens involved in the making of star catalogues has been the detection of DOUBLE STARS—stars that not only appear to be near together but are really so. Quite a distinct and important department of astronomy has arisen dealing with the continual observation and measurement of these objects. For many double stars are in motion round each other in obedience to the law of gravitation, and their orbits have been computed. Some of these systems contain three or even four members. But in every case the smaller body shines by its own light; we have no instance in these double stars of a sun attended by a planet; in each case it is a sun with a companion sun. The first double star to be observed as such was one of the seven stars of the Plough. It is the middle star in the Plough handle, and has a faint star near it that is visible to any ordinarily good sight.

Star catalogues and the work of preparing them have brought out another class—VARIABLE STARS. As the places of stars are not fixed, so neither are their brightnesses, and some change their brightness quickly, even as seen by the naked eye. One of these is called **Algol**, *i.e.* the Demon Star, and is in the constellation Perseus. The ancient Greeks divided all stars visible to the naked eye into six classes, or "**magnitudes**," according to their brightness, the brightest stars being said to be of the first magnitude, those not quite so bright of the second, and so on. Algol is then usually classed as a star of the second magnitude, and for two days and a half it retains its brightness unchanged. Then it begins to fade, and for four and a half hours its brightness declines, until two-thirds of it has gone. No further change takes place for about twenty minutes, after which the light begins to increase again, and in another four and a half hours it is as bright as ever, to go through the same changes again after another interval of two days and a half.

Algol is a double star, but, unlike those stars that we know under that name, the companion is dark, but is nearly as large as its sun, and is very close to it, moving round it in a little less than three days. At one point of its orbit it comes between Algol and the Earth, and Algol suffers, from our point of view, a partial eclipse.

There are many other cases of variable stars of this kind in which the variation is caused by a dark companion moving round the bright star, and eclipsing it once in each revolution; and the diameters and distances of some of these have been computed, showing that in some cases the two stars are almost in contact. In some instances the companion is a dull but not a dark star; it gives a certain amount of light. When this is the case there is a fall of light twice in the period—once when the fainter star partly eclipses the brighter, once when the brighter star partly eclipses the fainter.

But not all variable stars are of this kind. There is a star in the constellation Cetus which is sometimes of the second magnitude, at which brightness it may remain for about a fortnight. Then it will gradually diminish in brightness for nine or ten weeks, until it is lost to the unassisted sight, and after six months of invisibility it reappears and increases during another nine or ten weeks to another maximum. "Mira," *i.e.* wonderful star, as this variable is called, is about 1000 times as bright at maximum as at minimum, but some maxima are fainter than others; neither is the period of variation always the same. It is clear that variation of this kind cannot be caused by an eclipse, and though many theories have been suggested, the "**long-period variables**," of which Mira is the type, as yet remain without a complete explanation.

More remarkable still are the "NEW STARS"—stars that suddenly burst out into view, and then quickly fade away, as if a beacon out in the stellar depths had suddenly been fired. One of these suggested to Hipparchus the need for a catalogue of the stars; another, the so-called "Pilgrim Star," in the year 1572 was the means of fixing the attention of Tycho Brahe upon astronomy; a third in 1604 was observed and fully described by Kepler. The real meaning of these "new," or "temporary," stars was not understood until the spectroscope was applied to astronomy. They will therefore be treated in the volume of this series to be devoted to that subject. It need only be mentioned here that their appearance is evidently due to some kind of collision between celestial bodies, producing an enormous and instantaneous development of light and heat.

These New Stars do not occur in all parts of the heavens. Even a hasty glance at the sky will show that the stars are not equally scattered, but that a broad belt apparently made up of an immense number of very small stars divides them into two parts.

THE MILKY WAY, or GALAXY, as this belt is called, bridges the heavens at midnight, early in October, like an enormous arch, resting one foot on the horizon in the east, and the other in the west, and passing through the "**Zenith**," *i.e.* the point overhead. It is on this belt of small stars—on the Milky Way—that New Stars are most apt to break out.

The region of the Milky Way is richer in stars than are the heavens in general. But it varies itself also in richness in a remarkable degree. In some places the stars, as seen on some of the wonderful photographs taken by E. E. BARNARD, seem almost to form a continuous wall; in other places, close at hand, barren spots appear that look inky black by contrast. And the **Star Clusters**, stars evidently crowded together, are frequent in the Milky Way.

And yet again beside the stars the telescope reveals to us the NEBULÆ. Some of these are the Irregular Nebulæ—wide-stretching, cloudy, diffused masses of filmy light, like the Great Nebula in Orion. Others are faint but more defined objects, some of them with small circular discs, and looking like a very dim Uranus, or even like Saturn—that is to say, like a planet with a ring round its equator. This class are therefore known as "**Planetary Nebulæ**," and, when bright enough to show traces of colour, appear green or greenish blue.

These are, however, comparatively rare. Other of these faint, filmy objects are known as the "**White Nebulæ**," and are now counted by thousands. They affect the spiral form. Sometimes the spiral is seen fully presented; sometimes it is seen edgewise; sometimes more or less foreshortened, but in general the spiral character can be detected. And these White Nebulæ appear to shun the Galaxy as much as the Planetary Nebula; and Star Clusters prefer it; indeed the part of the northern heavens most remote from the Milky Way is simply crowded with them.

It can be by no accident or chance that in the vast edifice of the heavens objects of certain classes should crowd into the belt of the Milky Way, and other classes avoid it; it points to the whole forming a single growth, an essential unity. For there is but one belt in the heavens, like the Milky Way, a belt in which small stars, New Stars, and Planetary Nebulæ find their favourite home; and that belt encircles the entire heavens; and similarly that belt is the only region from which the White Nebulæ appear to be repelled. The Milky Way forms the foundation, the strong and buttressed wall of the celestial building; the White Nebulæ close in the roof of its dome.

And how vast may that structure be—how far is it from wall to wall?

That, as yet, we can only guess. But the stars whose distances we can measure, the stars whose drifting we can watch, almost infinitely distant as they are, carry us but a small part of the way. Still, from little hints gathered here and there, we are able to guess that, though the nearest star to us is nearly 300,000 times as far as the Sun, yet we must overpass the distance of that star 1000 times before we shall have reached the further confines of the Galaxy. Nor is the end in sight even there.

This is, in briefest outline, the Story of Astronomy. It has led us from a time when men were acquainted with only a few square miles of the Earth, and knew nothing of its size and shape, or of its relation to the moving lights which shone down from above, on to our present conception of our place in a universe of suns of which the vastness, glory, and complexity surpass our utmost powers of expression. The science began in the desire to use Sun, Moon, and stars as timekeepers, but as the exercise of ordered sight and ordered thought brought knowledge, knowledge began to be desired, not for any advantage it might bring, but for its own sake. And the pursuit itself has brought its own reward in that it has increased men's powers, and made them keener in observation, clearer in reasoning, surer in inference. The pursuit indeed knows no ending; the questions to be answered that lie before us are now more numerous than ever they have been, and the call of the heavens grows more insistent:

"LIFT UP YOUR EYES ON HIGH."

# INDEX